Digital Image Processing

Prentice Hall International Series in Acoustics, Speech and Signal Processing

Signals and Systems: An Introduction
L. Balmer

Signal Processing, Image Processing and Pattern Recognition
S. Banks

Randomized Signal Processing
I. Bilinskis and A. Mikelsons

Restoration of Lost Samples in Digital Signals
R. Veldhuis

Prentice Hall International Series in Acoustics, Speech and Signal Processing

Digital Image Processing

Jan Teuber
Copenhagen University Observatory

Translated from the Danish *Digital billedbehandling*

Prentice Hall
New York London Toronto Sydney Tokyo Singapore

First published 1993 by
Prentice Hall International (UK) Ltd
Campus 400, Maylands Avenue
Hemel Hempstead
Hertfordshire, HP2 7EZ
A division of
Simon & Schuster International Group

1st English edition Prentice Hall International (UK) Ltd, 1993

Typeset in 10/12 pt Times
by Mathematical Composition Setters Ltd, Salisbury, Wiltshire

Printed and bound in Great Britain
at the University Press, Cambridge

Library of Congress Cataloging-in-Publication Data

Teuber, Jan.
[Digital billedbehandling. English]
Digital image processing / Jan Teuber.
p. cm. – (Prentice Hall international series in acoustics, speech, and signal processing)
Translation of: Digital billedbehandling
Includes bibliographical references and index.
ISBN 0-13-213364-4 (pbk.)
1. Image processing–Digital techniques. 2. Signal processing–Digital techniques. I. Title. II. Series.
TA1632.T4813 1992
621.36′7–dc20 92-15478
CIP

British Library Cataloguing in Publication Data

A catalogue record for this book is available from the British Library

ISBN 0-13-213364-4 (pbk)

1 2 3 4 5 96 95 94 93

Contents

Preface

When looking back over the developments in society during the last two decades of scientific and technological advances, a picture emerges which is totally dominated by *data processing*. After the heyday of the pocket calculator through the 1970s, the 1980s were peppered by home computers and personal computers. With those, small miracles ensued, such as desktop publishing, facilitating a host of possibilities in the field of convenient graphical presentation of text and data.

It is to be foreseen that the last decade in our millennium similarly will be dominated by *digital image processing*. Two circumstances help substantiate this prophecy. First, the so-called *digital camera* (based on an integrated light detector known as a charge coupled device (CCD)) has recently been introduced. This impressive technical innovation would, however, be useless if it were not possible to store, transport, and process the enormous amount of data it produces. And since – secondly – the current trends in the computing world exhibit drastic expansions of storage, transmission, and processing capacity, there is a fertile soil nourishing the explosive growth witnessed in other areas of data processing.

This book is an attempt to anticipate this development. It should be stressed at once that it is solely devoted to the fundamental principles underlying digital image processing. Purely technical aspects (system realizations, software, etc.) or descriptions of commercially available products will only be treated very briefly; an extensive literature covering these subjects already exists. It is the maxim of this book that the very basic mathematical principles should be grasped before anything else. It is also stressed that the book does not assume access to a PC or image processing facilities as such.

The approach adopted in this book is to bring together certain theoretical topics which are normally treated separately. For instance, the reader will find no sharp distinction between images and other types of signal; on the contrary, a fairly comprehensive introduction to the subject of signal processing is provided, the only difference being the two dimensions of the new 'signals'. Moreover, I have strived to regard the concept 'analog' or 'continuous' as a derived concept, a limiting case of 'digital' – the reader should thus not expect a second volume entitled 'Analog

Image Processing'. Finally, the very general character of the concept of 'signal' or 'image' ensures that the presentation will be formally identical to that found in textbooks on linear algebra (manipulation of vectors = digital signals) or functional analysis (manipulation of functions = analog signals). This is, in my view, a great advantage.

The style of the book will probably be considered rather compact, but as the step from the fundamental principles to the practical applications is a very small one indeed, I hope that the illustrations and examples provided will suffice to guide the reader through the seven chapters with undiminished interest. The notation employed is also somewhat compressed. Thus, the same symbols are used for vectors, images, etc., as for ordinary numbers. It is my hope that this policy will highlight the fundamental principles hiding behind the calculations.

For good measure, it should be stressed that the presentation is a case of the usual compromise between, on the one hand, mathematical rigour and, on the other hand, intuition. The balance undoubtedly favours the latter aspect. A result being 'proved' will invariably rest upon a number of implicit assumptions; when an infinite series is being summed, for instance, its convergence is assumed.... So this is not a mathematical textbook – instead, I have tried to convey a solid foundation of ideas and principles with direct relevance for image processing in theory and practice.

The examples encountered during reading should be considered an integral part of the presentation, and the reader is urged to stop and work on them before proceeding.

After Chapter 1, furnishing some background information, we move through the topics of representation, transmission, and processing of image information. Image processing *per se* is relegated to Chapters 6 and 7, where all the tools and techniques touched upon previously are applied.

The reader will surely not fail to notice that the author's point of view is that of astronomy. It is, however, with a clear purpose in mind that this subject makes its presence felt in the illustrations. The reason is that astronomical images provide an ideal ground for approaching the principles of image processing within a fully realistic framework (see, for example, p. 106).

Another word about the photographic illustrations. It is ironic that raster-based printing techniques, with their rather poor reproduction of grey levels, provide the only ones that are economically reasonable! The reader is thus, regrettably, only able to use the illustrations as intuitive guides – it is out of the question to measure grey levels and check the author's veracity, as one would do with a mathematical proof. More or less in desperation, I have resorted in several places to 'illustrations of principle' where I feel that the slightly humorous touch will have some pedagogical effect. I can assure the reader that the impression conveyed from a glance at the illustrations in question – the only available option in the absence of the digitized values themselves – will be perfectly adequate.

My thanks are due to Karsten Holten, the indefatigable editor at Teknisk Forlag, Copenhagen, for valuable advice and support. Many of the computer screen images have been prepared with the assistance of Dr Leif Toudal, whom I wish to thank warmly.

Jan Teuber
Copenhagen, November 1991

1

Introduction

1.1 Digital quantities

The term *digital* has become part of our daily lives. It derives from Latin *digitus*, finger, and has since then quite understandably come to mean 'numerical' or 'number-related'. The word is known from constructs such as 'digital clock' (a clock showing time as numbers or digits); even 'digital sound' is a well-known concept today.

What characterizes digital quantities is their being determined with a certain, necessarily finite, number of digits – hence, a limited degree of accuracy. Any quantity to be specified, that is, conveyed in writing or by some other means, will accordingly be digital; an infinite number of digits is, of course, impossible. But we can say more than that.

The number of digits to appear can be chosen in an optimal way. If someone asks me the time, I have been taught to be polite, so I do not answer '1992' or '21st March'; on the other hand, I do not answer '496 thousandths of a second past 6.28' either. The display on a digital clock – typically, hours, minutes, and seconds – does not solely contain the solid knowledge of the point of time in question; it also implies a choice of an appropriate level of information corresponding to our daily needs, so that we are neither drowning in unusable data nor in short supply of information.

The choice of number of digits may also be dictated by purely technical considerations. If my digital clock gains a couple of minutes per hour, and if I am unable to correct it, then its seconds display will be utterly meaningless. As a consequence, this clock should display four digits (hours and minutes) – no more, no less.

Another advantage of digital techniques, quite often referred to but nearly as often misinterpreted, is the elimination of *noise*. When transmitting data, for instance when answering the question 'What time is it?', there *always* exists a possibility that the information will be conveyed incorrectly; even '6.28' may easily become '8.26'. The essential difference in transmitting digital quantities is the constancy of the fundamental information unit to be transferred; normally, this unit is referred to as one *bit* (*b*inary dig*it*), denoted by either 0 or 1. Accordingly, it will be possible to acquire a substantial amount of knowledge about the mechanisms causing the noise and to design the transmission systems accordingly, so that the noise will have a tolerable and, above all, a *well-defined* magnitude.

1.2 Digital images

A *digital image* is an image characterized by means of numbers. How can this characterization be accomplished in, for example, the black and white photograph in Figure 1.1? The answer will become clear from Figure 1.2, where sections from Figure 1.1 have been successively enlarged. The sequence illustrates that the original image does not contain unlimited amounts of information – it may only be magnified to a certain point, beyond which no further details will appear. Quite generally, one will tend to present one's pictures in a form rendering continued magnification pointless.

The subimages thus rapidly lose their identities as (recognizable) parts of a 'proper' image, and we note three distinct types: subimages with continuous variations of grey levels (Figures 1.2a and 1.2b); those with approximately constant density of grey (Figures 1.2c and 1.2d); and those consisting of individual photographic grains (Figures 1.2e and 1.2f). Continued magnification will, of course, result in either 0 or 1 grain per subimage everywhere, if we disregard the size of the grains themselves.

Thus the *number* enters in the world of images. The original image is divided into fields, and the number of grains in each field is specified. (For clarity, this

Figure 1.1 Portrait of the author as a young man.

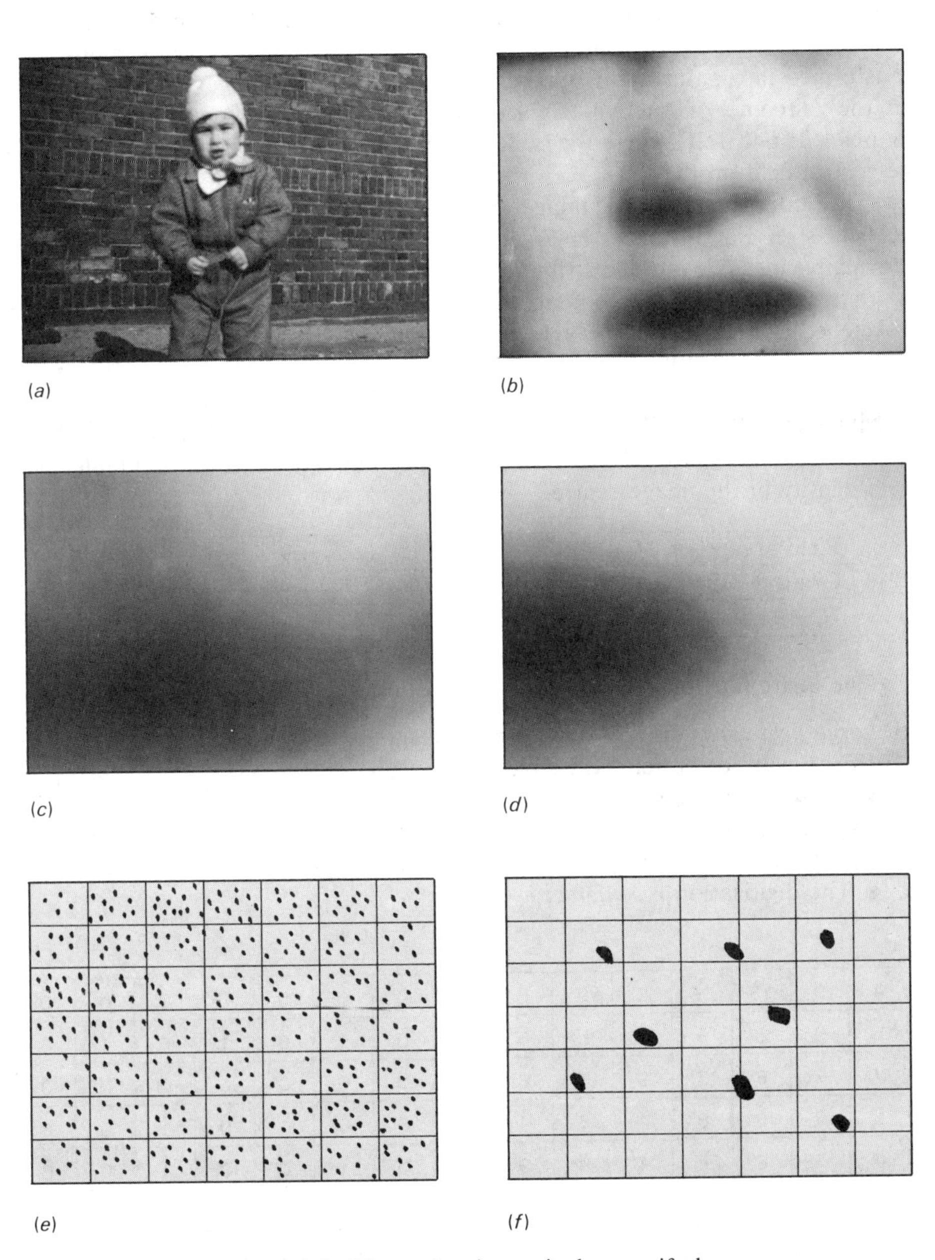

(a) (b) (c) (d) (e) (f)

Figure 1.2 The author increasingly magnified.

division has only been carried out in Figures 1.2e and 1.2f). The choice of rectangular or, more often, square fields, logically oriented as the original one, is normal practice everywhere in digital image processing. The image has now been transformed into an array of numbers – our first digital image. In Figure 1.3, we see the outcome of such a *digitization process* applied to the subimages in Figures 1.2e and 1.2f, where the table to the right is the result of magnification of the central field on the left.

The 'looks' of a digitized image, that is, its numbers, will depend strongly upon the field size. For instance, a larger area should contain more grains. A more informative measure of the distribution is, then, the *density* of grains – the number of grains per unit area in the image. The variation of this quantity with field area is sketched in Figure 1.4. The field centre is assumed to be fixed.

EXAMPLE 1.1

Use the image in Figure 1.3b as a unit area and complete the following table giving grain density in the image centre.

Field area	1/49	9/49	25/49	1	9	25	49
Grain density							

The figure illustrates, once again, the conditions (a)–(c) above:

(a) *Large field areas* display variations corresponding to the image itself, and the density will not be representative.
- The digitization is *too coarse.*

(b) For a certain range of areas, the density is *independent of area*. This density is exactly the relevant one for the image point in question.
- The digitization is *satisfactory*.

4	9	13	10	8	9	5
6	9	8	5	7	6	5
10	7	6	7	9	7	8
6	8	6	8	4	6	4
4	7	6	8	4	8	9
8	7	7	6	10	11	10
5	7	9	6	8	7	7

(a)

0	0	0	0	0	0	0
0	1	0	1	0	1	0
0	0	0	0	0	0	0
0	0	1	0	1	0	0
0	1	0	0	1	0	0
0	0	0	0	0	1	0
0	0	0	0	0	0	0

(b)

Figure 1.3 Digitized portions of the author.

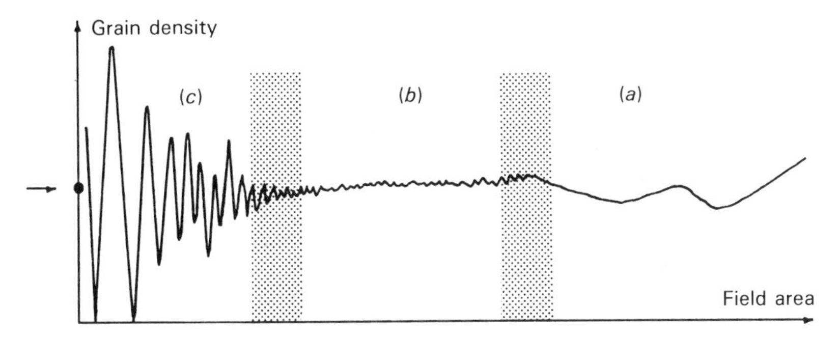

Figure 1.4 Grain density at a fixed point in the image.

(c) *Small field areas* only contain few grains, and the density fluctuates violently.
- The digitization is *too fine.*

Only the value from range (b) merits the designation *the* grain density (corresponding to the image point under consideration); it has been marked with an arrow on the vertical axis. In the other intervals, one must be content with specifying the *average* grain density.

The grain density is the quantity responsible for the eye's perception of grey level. The eye is able to simulate the counting process described above and is said to act as an *integrator* or *lowpass filter* (Section 7.2) under these circumstances.

The number of photographic grains is, on the other hand, a measure of the amount of light incident upon the area in question during the total exposure time. The grain density thus also records the *intensity* of the light, i.e. the amount of energy received per unit time and per unit area. In order to speak of *the* intensity, this quantity must be assumed constant during exposure, just as was the case for the grain density, regarded as a function of area.

Light intensity is measured in units of W/m^2 (watts per square metre). The unit W represents the physical quantity called *power*, i.e. energy per unit time, and is equal to J/s (joules per second). The unit for intensity is, consequently,

$$W/m^2 = J/s/m^2 = J\,s^{-1}\,m^{-2}$$

Summing up, we note the very close relation between the following three quantities characterizing a definite position in a photographic black/white image:

1. the digital quantity *grain density*;
2. the physiological quantity *grey level*;
3. the physical quantity *light intensity*.

For completeness' sake, it should be noted that the above considerations only apply to photographic negatives – grains are, unfortunately, black!

1.2.1 Digitization of images

The process described above, where an image is divided into small fields, each of which is assigned a value for its grain density, its grey level, or its light intensity, is known as *digitization*. Obviously, many ways for doing this will exist, as one is free to choose, at least in principle, between various divisions into fields and various specifications of the number assigned to the field. The process is, nevertheless, reasonably unambiguous.

The *field size* should, as shown above, be chosen so as to ensure an area-independent and thus well-defined intensity. In other words, the grey level should look uniform over each field. For obvious practical reasons, one would normally employ the same field size for the entire image, and the preferred size should thus accommodate all the image points at the same time – not only the randomly selected point giving rise to Figure 1.4. The freedom suggested from this illustration, in the range (b), is thus somewhat restricted.

The *number* appearing in the field may be chosen as the actual number of physical grains or, perhaps, in a simpler way which represents these actual numbers adequately. For instance, the human eye is only able to distinguish between a limited number of grey levels. It may, accordingly, be decided to employ a 'code' with this number of 'code words'. This coding process is called *quantization*. One will often encounter the poorly chosen term 'A/D conversion' (A/D: analog/digital) or, worse still, 'digitization'.

EXAMPLE 1.2

When digitizing a certain image, one finds grain numbers varying from 8128 to 11 941. The image is to be displayed on a computer screen possessing 64 grey levels, represented by the integers from 0 to 63. Employing a straightforward (linear) quantization, the grain number 10 000 will be coded as

$$K = \text{int}\left(\frac{10\,000 - 8128}{11\,941 - 8128} \cdot 63\right) = 30$$

where int stands for 'the integral part of'. What are the quantization values for the physical grain numbers 9000 and 11 000?

The practical digitization of photographic images is a rather complicated process. Where the demands for quality have not been prohibitive, television or video cameras have been the natural tools. The output signal, normally consisting of line scans arranged sequentially, is then quantized electronically over small time intervals. Also, various types of scanning devices – which look like copying machines – exist. For more professional use, the so-called *microdensitometer* has been utilized; here, a thin light ray, for instance a laser beam, is passed through a transparent film or plate containing the image. The cross-section of the beam corresponds to the digitizing area, and the grey level is measured from the intensity decrement caused by the passage.

The recent and seemingly universal solution of the problem concerning an efficient and reliable digitization is probably the *digital camera*, based upon a light-sensitive semiconductor unit known as a CCD chip (see Section 1.4). By means of this device, existing images may be conveniently digitized, whereas new images may be *recorded* digitally, that is, without ever having been represented in the usual physical form.

How is an image rendered in its 'normal' appearance (for example, on a computer screen), starting from a digitized version? First, the available physical display area is divided into a number of fields corresponding to the actual digitization. Next, each field is given a grey tone value determined by the number assigned to the field in question. This process of reconstruction is illustrated in Figures 1.5 and 1.6. The original image and its reconstructed versions are shown for various field areas (Figure 1.5) and different numbers of quantization levels (Figure 1.6).

In view of the *image quality*, it is, of course, desirable to choose the largest possible number of fields and quantization levels, especially if unlimited data storage and processing resources are available. Alas, life is never that simple, and one's choice will inevitably be a compromise between quality and capacity.

EXAMPLE 1.3

If one is content to retain 'what the image looks like', the quality represented by Figure 1.5d (128×96 fields) and Figure 1.6e (4 levels) seems adequate. The best quality corresponds, however, to 512×384 fields and 64 levels. The possibly acceptable quality deterioration thus implies a demand on capacity slackened by a factor of $(512 \times 384 \times 64)/(128 \times 96 \times 4) = 256$.

EXAMPLE 1.4

A computer's storage capacity is measured in 'bytes' (1 byte = 8 bits), abbreviated b. Storage of 2^n levels requires n bits. Our computer has an available memory of 600 000 b. How many images, represented with the highest quality as in Example 1.3, will it hold?

The digital characterization of an image thus involves a range of values giving the intensities at all the image points. Also, the position of these points in the image must be specified. This is done by means of a *coordinate system*.

In this book, we shall exclusively deal with rectangular images, and the coordinate systems to be used will be placed as shown in Figure 1.7. The origin is placed either in the lower left-hand corner or at the image centre. In practical

(a) Original

(b) 512 × 384 fields

(c) 256 × 192 fields

(d) 128 × 96 fields

(e) 64 × 48 fields

(f) 32 × 24 fields

Figure 1.5 Different choices of digitization area.

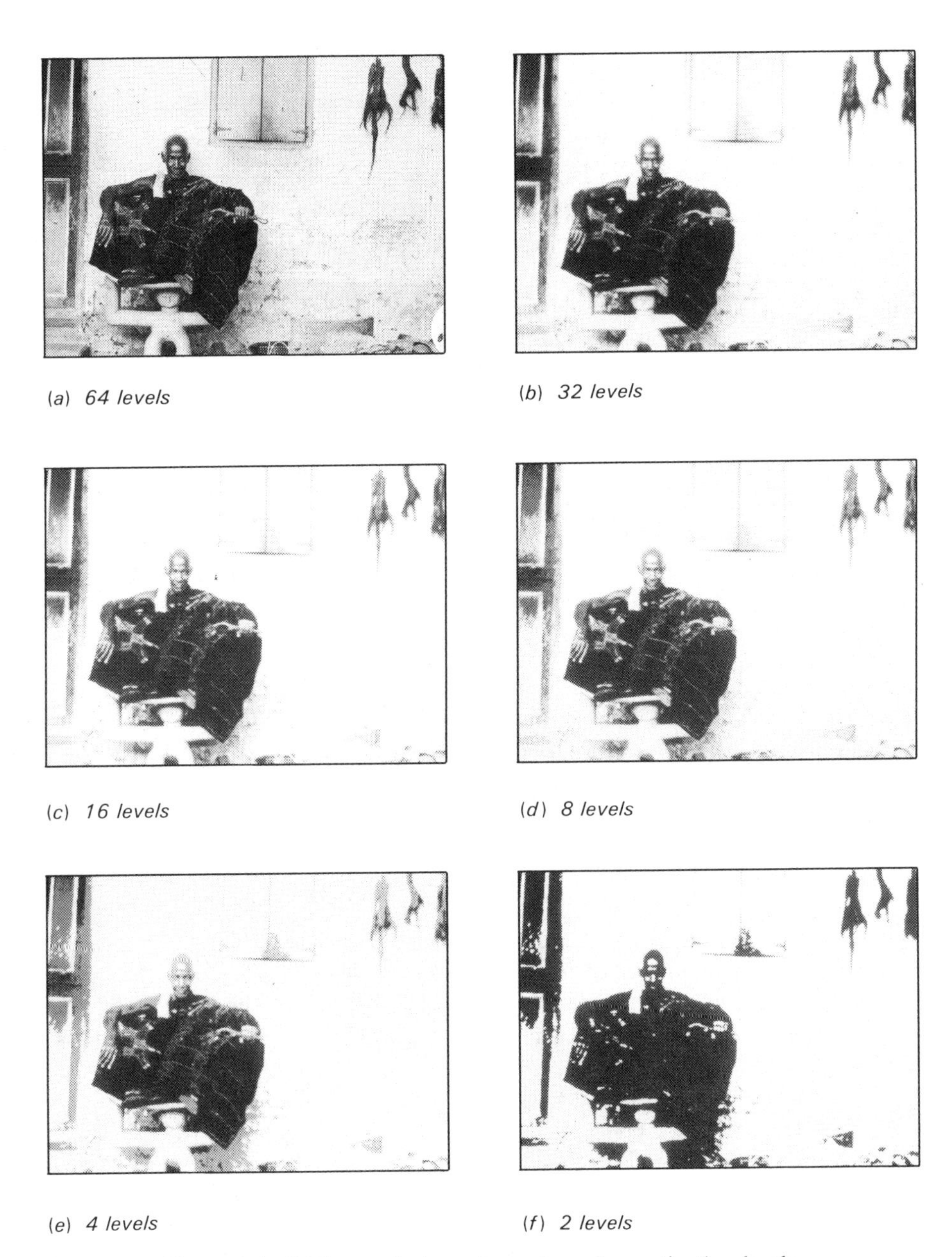

(a) 64 levels
(b) 32 levels
(c) 16 levels
(d) 8 levels
(e) 4 levels
(f) 2 levels

Figure 1.6 Different choices of number of quantization levels.

Figure 1.7 Image with two coordinate systems.

applications, one often encounters coordinate systems with origin in the upper left-hand corner of the image and the y-axis pointing vertically downwards.

Having made these preparations, a digital image is now simply a function of two variables:

$$b = b(x, y)$$

expressing the grain density (or light intensity) b at the point (x, y).

There is, however, no reason to limit the scope to photographs – *all* images are digital! The reasons are as follows. First, a position in an image may only be specified with a certain degree of accuracy. An image point with coordinates (3.5 cm, 7.9 cm) is, of course, not a perfect mathematical point; in the format of this specification, one is implicitly induced to think of a small field of dimensions 0.1×0.1 cm, in principle centred upon (3.5000... cm, 7.9000... cm). This fact is reflected in similar physical limitations. The *human eye*, for example, perceives the incoming light via a number of discrete receptory units (the 'rods' and 'cones' in the retina, see Section 1.4), and changes in physical stimulus over scales smaller than the corresponding receptor size will not be recorded. In a TV set, the physical dimension of the electron ray hitting the screen limits the spatial resolution. A CCD chip is *constructed* as an array of rectangular units. The smallest physical unit in an image able to record, represent, or produce a discrete amount of light is called a *picture element* or *pixel*. The pixel size defines the spatial resolution of the image.

Secondly, light is *propagated* as electromagnetic waves (see the next section). Light is, however, always *detected* in separate portions, light *quanta* or *photons*, each of which represents a definite amount of energy. This unit or 'package' of energy E is proportional to the frequency ν of the light:

$$E = h\nu \tag{1.1}$$

The constant of proportionality is called *Planck's constant* after the German physicist[1] who discovered this relationship in 1901. Its value is

$$h = 6.626 \times 10^{-34}\ \mathrm{J\,s}$$

The reader is reminded that the frequency ν and the *wavelength* λ are inversely proportional:

$$\lambda\nu = c \tag{1.2}$$

where c is the *velocity of light*:

$$c = 2.998 \times 10^{8}\ \mathrm{m\,s^{-1}}$$

that is, nearly 300 000 kilometres per second.

EXAMPLE 1.5
A light quantum with the wavelength $\lambda = 500$ nm (1 nm = 1 nanometre = 10^{-9} m) has a frequency $\nu = 6.00 \times 10^{14}$ Hz and carries an energy $E = h\nu = 4 \times 10^{-19}$ J.

The amount of light associated with a certain pixel in a given image hence corresponds to a certain number of light quanta. In summary: as a consequence of the limited spatial resolution inherent in every physical system and the nature of light as discrete quanta, every image is, in principle, a digital one.

Each processed grain in a photographic emulsion may be regarded as a detection of one photon. If the opposite were true, that is, if every photon incident upon the emulsion were to result in a grain, we would possess an *ideal* two-dimensional (also called *panoramic*) light detector. Within a given area, the number of grains would equal the number of photons, and the total light energy could be obtained by counting (and multiplication by an appropriately chosen mean energy corresponding to the distribution of wavelengths). In the real world, only a certain *fraction* of the photons will lead to the creation of a grain. This fraction is called the *quantum efficiency*. For photographic emulsions, it is of the order a few per cent.

EXAMPLE 1.6
On a 1 cm^2 area of a photographic film, an exposure of 15 seconds has triggered 5×10^5 grains. The quantum efficiency is specified to be 0.02, and the wavelength of the light in question is 550 nm. Find the intensity in W m^{-2}.

---

[1] Max Planck (1858–1947)

For the CCD chip, the quantum efficiency may exceed 90 per cent, even over a wide wavelength range. This detector type is, accordingly, often referred to as being close to ideal.

1.3 Properties of light

1.3.1 Light and other electromagnetic radiation

Light is a form of electromagnetic radiation, consisting of vibrations propagating through empty space with a velocity of nearly 300 000 km s^{-1}, as mentioned above. Light is distinguished from the other electromagnetic waves in being *visible*.

Visible light is characterized by its wavelength λ (or frequency ν, cf. equation (1.2)) in the range from about 400 nm to 700 nm. Wavelength is perceived as *colour*. Table 1.1 gives the approximate wavelength corresponding to light of a given colour. Light occurring normally is a mixture of oscillations with a wide range of wavelengths, and the colour perceived is determined by the 'mixing fractions'. For instance, light from the Sun has a wavelength distribution as depicted in Figure 1.8. As the figure shows, the human eye is sensitive to exactly those wavelengths characterizing solar radiation – a fact explained by evolutionary adaptation.

Table 1.1 The relationship between wavelength of light and colour.

Colour	red	orange	yellow	green	blue	violet
Wavelength (nm)	675	625	575	525	475	425

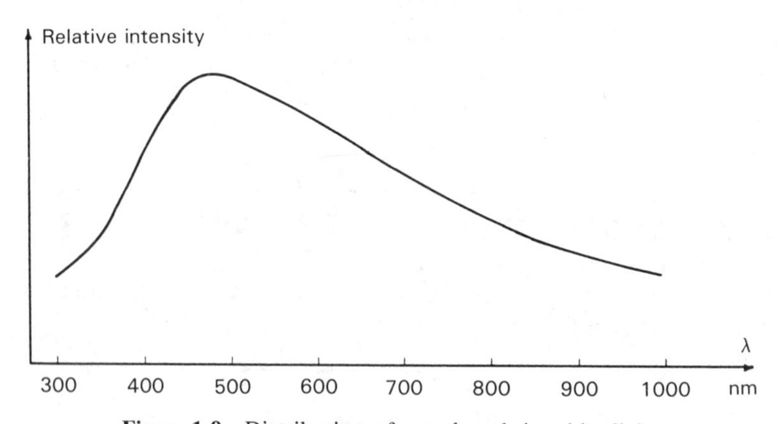

Figure 1.8 Distribution of wavelength in white light.

The other, invisible, electromagnetic waves with wavelengths below 400 nm or above 700 nm are listed in Figure 1.9.

There are no other differences between the radiation types (all electromagnetic waves, for example, travel through space at the speed of light). The different designations are due to the very different physical processes producing the radiation, and the equipment required for its detection and measurement.

1.3.2 Propagation of light

The path of a light ray in a vacuum is a straight line. The propagation of the light ray causes a 'disturbance' of space in the sense that, along the ray, a combination of an electric and a magnetic field is produced.

If we examine the electric field along the path, we find the variation shown in Figure 1.10. The electric field vectors are always perpendicular to the direction of propagation. They are, however, not necessarily located in the same plane, as indicated in the figure (the plane defined by the page).

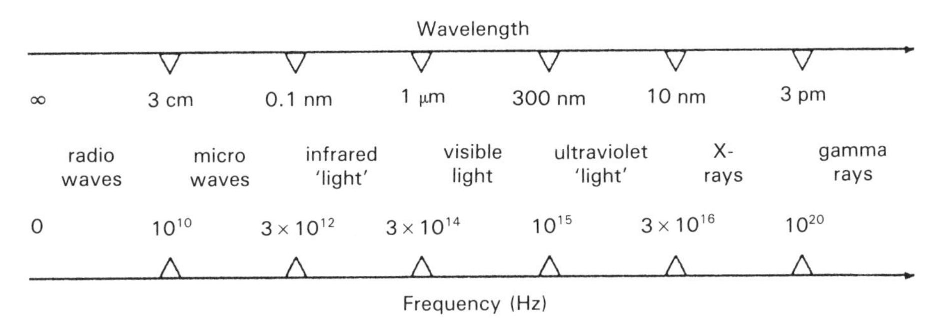

Figure 1.9 The electromagnetic radiation spectrum.

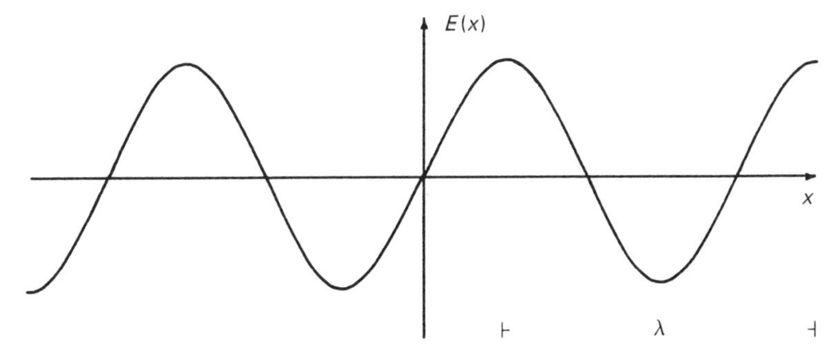

Figure 1.10 Simultaneous values of the electric field.

With an appropriately chosen zero point for the path coordinate x, the variation of the electric field E is given by

$$E = E_0 \sin kx \tag{1.3}$$

where the two constants E_0 and k are known as the *amplitude* and *wavenumber*, respectively, for the field E. This (trigonometric) relationship is called *harmonic*; thus, the electric field varies harmonically with position in space.

The amplitude may be characterized as the maximal value of the field. The wavenumber is related to the wavelength λ of the light, being defined as the distance corresponding to a full period of oscillation:

$$k\lambda = 2\pi$$

that is,

$$k = \frac{2\pi}{\lambda} \tag{1.4}$$

The situation in Figure 1.10 describes a snapshot of the field. If we next turn to the temporal variation at a fixed point in space, we find the behaviour shown in Figure 1.11, and the time dependence for the field at this point – and at all other points along the ray – is also of the harmonic type:

$$E = E_0 \sin \omega t \tag{1.5}$$

The amplitude of the field is E_0, the same quantity as above. However, the *(angular) frequency* ω is linked to the *oscillation period* T, that is, the time interval corresponding to one full oscillation:

$$\omega T = 2\pi$$

that is,

$$\omega = \frac{2\pi}{T} \tag{1.6}$$

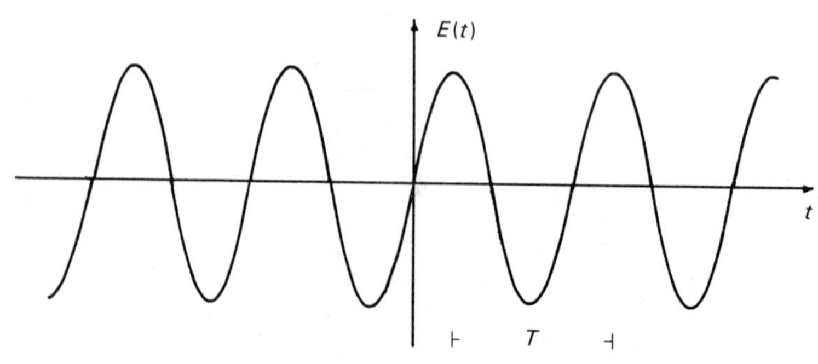

Figure 1.11 The variation of the electric field at a fixed point.

It is now quite easy to express the complete dependence of the E-field, $E(x,t)$, as a function of the path coordinate x and the time t. The argument runs as follows.

Equation (1.3) for the field described the situation at a given instant in time, the only assumption being the choice of zero point for x at a position with field zero. For different instants t, one must necessarily find the same relation *except for the zero point*:

$$E(x,t) = E_0 \sin k(x - x_0(t)) \tag{1.7}$$

Analogously, we can compare the temporal variation (equation (1.5)) for different positions x:

$$E(x,t) = E_0 \sin \omega(t - t_0(x)) \tag{1.8}$$

The two latter expressions must be identical for all choices of x and t. Thus,

$$kx - kx_0(t) = \omega t - \omega t_0(x)$$

If this equation, however, is rewritten as

$$kx + \omega t_0(x) = kx_0(t) + \omega t$$

it states that a certain function of x (the left-hand side) is identical to a certain other function of t (the right-hand side). This is, however, only possible if both functions are constant, that is, independent of position and time:

$$kx + \omega t_0(x) = kx_0(t) + \omega t = \phi$$

On substitution in equation (1.7) or (1.8), one finally finds

$$E(x,t) = E_0 \sin(kx + \omega t - \phi) \tag{1.9}$$

This expression for the field corresponds to a 'running wave' – one propagating with unaltered shape in the negative x-direction.

To determine its speed, we note that the passage of the time interval Δt is 'compensated' by a spatial translation Δx as follows:

$$kx + \omega t = k(x + \Delta x) + \omega(t + \Delta t)$$

if one chooses

$$\Delta x = -\frac{\omega}{k}\Delta t$$

In other words, any point characterized by a given field strength will move through space with velocity

$$c = \frac{\omega}{k} \tag{1.10}$$

in the negative x-direction.

EXAMPLE 1.7
The frequency ν and the period T are reciprocal quantities. From this, it follows (with due reference to equation (1.10) as well as equations (1.4) and (1.6)) that

$$c = \lambda \nu$$

as stated earlier (equation (1.2)).

The harmonic variations, of which the electrical field in equations (1.3), (1.5), and (1.9) provides an example, will play a fundamental role throughout this book. The independent variable (kx, ωt, or the combination $kx + \omega t - \phi$) is the field's *phase angle* or just *phase*.

For harmonic functions, the so-called *complex notation* is often used. This notation is based upon the relationship between trigonometric functions and complex exponentials:

$$e^{i\alpha} = \cos \alpha + i \sin \alpha$$

In particular, $\cos \alpha$ is given by the *real part* of $e^{i\alpha}$ (denoted Re):

$$\cos \alpha = \text{Re}(e^{i\alpha})$$

This relationship implies a recipe for converting from trigonometric to complex notation. As the latter is far more convenient for practical calculations, it will be employed systematically in what follows. Where a real quantity appears on one side of an equation and a complex quantity on the other side, it is implicit that we are taking the real part of the latter.

EXAMPLE 1.8
In complex notation, equation (1.9) is rewritten as

$$\begin{aligned} E(x, t) &= E_0 \sin(kx + \omega t - \phi) \\ &= E_0 \cos(\pi/2 - (kx + \omega t - \phi)) \\ &= E_0 e^{i((kx + \omega t) - (\phi + \pi/2))} \\ &= A e^{i(kx + \omega t)} \end{aligned}$$

where the notation is further compressed by allowing the amplitude to assume complex values, too:

$$A = E_0 e^{-i(\phi + \pi/2)}$$

The complex amplitude will often be chosen so as to absorb possible constants from the phase. Its modulus $|A|$ is, however, equal to the original, real amplitude, as before.

EXAMPLE 1.9
Demonstrate, using complex notation, that

$$\sum_{n=0}^{N-1} \cos(n\alpha) = \tfrac{1}{2} + \frac{\cos(N-1)\alpha - \cos N\alpha}{2(1 - \cos \alpha)}$$

The type of wave disturbance discussed here is called a *monochromatic* wave, as it is described by means of a single wavelength. A very important property of electromagnetic fields is that they obey the *superposition principle*: if two or more fields $E_1, E_2, \ldots$ exist separately, any *linear combination* $\Sigma\, A_n E_n$ is allowed as well. If, for the moment, we see fit to ignore that the field is a vector quantity, the most general form of such a linear combination will be

$$E(x, t) = \int A(\lambda) e^{i(2\pi/\lambda)(x-ct)}\, d\lambda \tag{1.11}$$

This calls for a little explanation. The *integral sign* just emphasizes the fact that many wave components are present, each one weighted by its 'density'. The density is specified by the complex amplitude $A(\lambda)$, now a function of wavelength. Finally, in the exponent we have utilized equations (1.4) and (1.10), as well as the fact that the propagation speed c is independent of wavelength.

In the case of a light ray from the Sun, $|A(\lambda)|^2$ varies as shown in Figure 1.8.

For completeness, we finally note that all the above considerations pertaining to the electric field apply to the magnetic field as well. In a *linearly polarized* wave, the electric field vectors are all found in a common plane which, of course, contains the direction of propagation. The same may be said of the magnetic field B, but the two planes in question are perpendicular to each other, as shown in Figure 1.12.

The use of polarizing filters allows the separation of the two perpendicular components. In Figure 1.13, a water surface has been photographed twice through a polarizing filter; the orientation of the electric field vector is indicated by the arrows. It follows that the electric field in the reflected light is preferentially polarized parallel to the water surface.

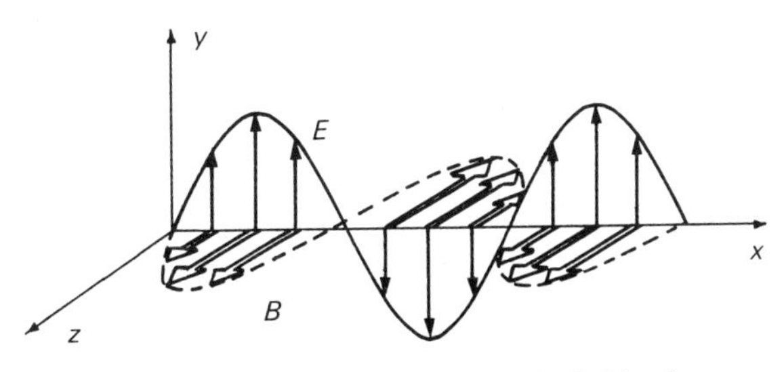

Figure 1.12 The electric and magnetic fields along an electromagnetic wave.

Figure 1.13 A lake viewed through a polarizing filter.

1.3.3 Detection and measurement of light

As mentioned above, light waves carry energy. If a point source emits a certain amount of energy in the form of light, and if this energy – prior to detection – is collected in an area D_1 at a distance r_1 as shown in Figure 1.14, then the same amount of energy could be collected at the larger distance r_2, provided that the area, D_2, available at this larger distance was

$$D_2 = \left(\frac{r_2}{r_1}\right)^2 D_1$$

The intensity I, that is, the energy received per unit time and unit area, is thus inversely proportional to the square of the distance to the source:

$$I \propto r^{-2} \tag{1.12}$$

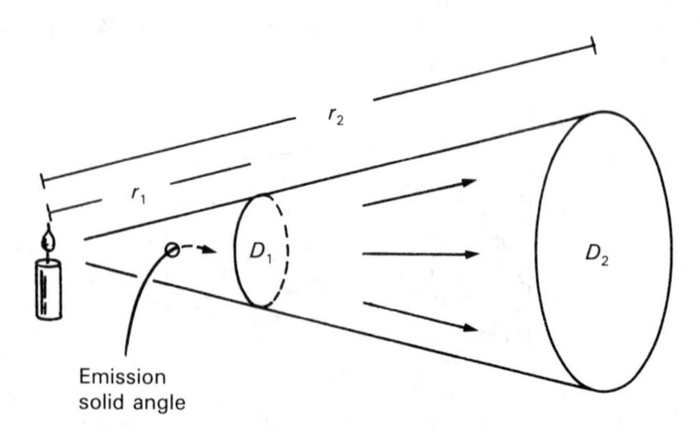

Figure 1.14 Detection of light from a point source.

As can be shown theoretically, the intensity of an electromagnetic wave is proportional to the square of the amplitude:

$$I \propto |A|^2 \tag{1.13}$$

From expressions (1.12) and (1.13) it follows that the amplitude decreases inversely with distance from the source:

$$|A| \propto r^{-1} \tag{1.14}$$

Returning for a moment to Figure 1.10, we observe that the variation of the field, as illustrated, corresponds to a *distant* source, so that $|A|$ is constant over the portion of the wave being considered. However, the temporal variation in Figure 1.11 is correct, regardless of the distance of the source, provided it is constant in time.

In image processing, sources are never points but extended objects seemingly appearing on a plane positioned at right angles to the line of sight. If we imagine this plane as being coated with point sources (Figure 1.15), the light from which is collected in one and the same area, then we may – by restricting ourselves to small object areas – assume a *homogeneous* coating of point sources (cf. also Figures 1.2c and 1.2d). The intensity of light received from the object plane will in this case be proportional to the *object area* considered, and the relevant quantity is now the light intensity per object surface area. As one rarely is granted 'direct access' to the object, a far more useful quantity is the *intensity per (reception) solid angle*. This quantity, which is of fundamental importance throughout radiation and image analysis, is called the *surface brightness* of the source and is denoted b:

$$\begin{aligned} b &= \text{intensity per reception solid angle} \\ &= \text{(received) radiation energy/time} \times \text{area} \times \text{solid angle} \end{aligned}$$

The surface brightness possesses the rather surprising property of being independent of the distance between light emitter and receiver. This derives from the fact that the intensity of light from each point source and, consequently, each small object

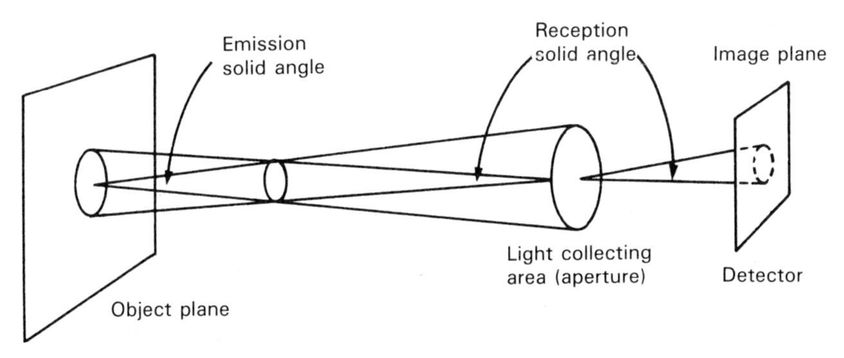

Figure 1.15 Detection of light from an extended source.

area, decreases with distance according to expression (1.12), but the reception solid angle for the object area decreases in exactly the same manner. This is expressed in the following statement:

The surface brightness is constant along the line of sight.

The surface brightness was introduced as a quantity characterizing the local radiation field, but it is nevertheless exactly equal to what is measured in the immediate vicinity of the source – a surprising result in view of the fact that modern telescopes are able to penetrate billions of light years outwards in space.

The surface brightness will, in general, vary with position in the object plane and should thus be considered a function b of two variables specifying this position. The two variables may be geometric coordinates associated with the source, but a much more useful strategy is to specify the *direction of the line of sight* to the object position. This is almost always achieved by means of a plane mapping of the object, mostly onto a detector plane (photographic film, the retina of the eye, a CCD chip, etc.). In this plane mapping, *a given detector area corresponds to a certain reception solid angle*, cf. Figure 1.16. This means that the intensity across the *detector area* will be a direct measure of the surface brightness of the *source*. The latter quantity may thus be added to the list on p. 5 as item 4). This justifies the use of the same notation b for digital images (p. 10) and surface brightness (p. 19).

In summary

A digital image of an object is a numerical specification of the surface brightness of its light. This specification is effected by means of a function of two variables:

$$b = b(x, y)$$

where (x, y) are coordinates for position within the image, that is, the direction to the corresponding object point.

What has been said so far only goes, strictly speaking, for black and white images. We have been thinking of light as being composed of colours according to a certain distribution. The intensity values will, of course, depend upon this distribution and the colour sensitivity of the detector system.

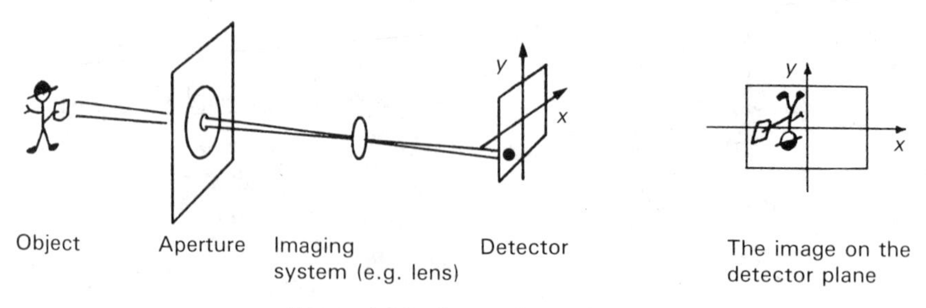

Figure 1.16 Image formation.

1.4 Two important light detectors

In this section, we describe two efficient light detectors: the human eye and the modern CCD camera. Both are two-dimensional detectors – the detection takes place over physical areas. But there all similarities end. It is to be foreseen, however, that these two detectors will dominate the subject of light detection in the future, and it is for this reason that other detector types – including photographic techniques – are not mentioned below.

1.4.1 The eye and the visual system

The human eye, one of nature's most highly specialized sensory units, is shown in cross-section in Figure 1.17. The eye is almost spherical, with a diameter of about 2.5 cm. It is encapsulated by three layers: the outermost being the *sclera*, the foremost part of which is the transparent *cornea*; the middle layer being the *chorioidea*; and the innermost layer being the *retina*. Muscles control both the shape of the *lens*, which determines the focal length of the eye, and the size of the pupil, regulating the total amount of light entering the eye. The former activity is called *accommodation*.

The lens of the eye forms an inverted image of the *field of view* upon the retina. This image is recorded by the *receptors* in the retina, known as *rods* and *cones*. These receptors are responsible for two quite distinct visual functions. The rods are active at low light levels (night or *scotopic* vision). The number of rods distributed across the retina is of the order of 100 million, but as the rods are connected in groups to a common nerve end, their spatial resolution is rather poor. The cones are responsible for our day or *photopic* vision, and the perception of colour is associated with these receptors. There are about 6–7 million of these, and they are connected to one nerve end each, resulting in high spatial resolution.

The distribution of receptors across the retina is by no means uniform, as shown in Figure 1.18. As will be noted, the cones are predominant in a small region

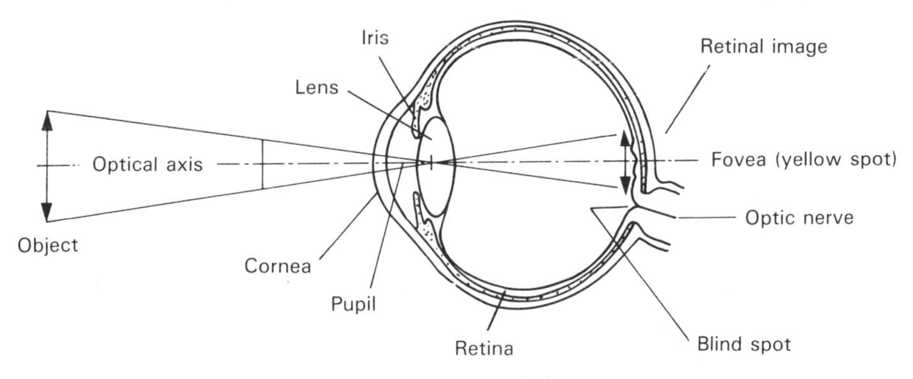

Figure 1.17 Cross-section of the human eye.

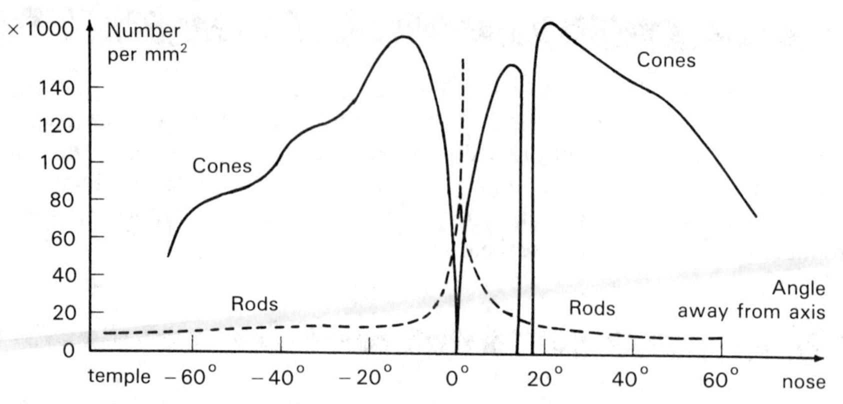

Figure 1.18 Distribution of rods and cones across the retina.

called the *yellow spot* or *fovea*, near the optical axis of the eye. In another small region, the so-called *blind spot*, no receptors are found; this is where the individual nerves are collected into the *optical nerve* passing the signals on to the visual cortex in the brain for processing.

As noted earlier, our vision is closely related to solar light. Thus, the sensitivity of the cones roughly reflects the distribution shown in Figure 1.8, whereas the response curve for the rods is slightly displaced towards the left (the blue, short-wavelength end of the spectrum).

It is possible to measure, more or less quantitatively, the physiological response corresponding to a given physical stimulus. One finds that the *sensory signal is proportional to the logarithm of the stimulus* (Weber–Fechner law). This is the reason why a logarithmic measure – similar to the one employed in acoustics – is employed in the case of light intensities:

$$s - s_0 = 10 \log_{10}(I/I_0) \text{ dB}$$

The physical stimulus, that is, the intensity I, is measured relative to a reference intensity I_0. Its value may be given as the above difference, in dB (decibels).

EXAMPLE 1.10

When going from the brightest stars in the night sky to the faintest ones, an approximate intensity ratio of 100 is found. What is this ratio in dB?

The eye responds to light intensities with a range of about ten orders of magnitude (relative values from 1 to 10^{10}), that is, a 100 dB scale. One should not be misled into believing that this range may be represented in one visual image. On the contrary, the eye *adapts* to a certain average intensity level for a given field of view.

The amount of light treated as a single image, a snapshot, by the visual system is set by a number of factors, including the pupil area (about 10 mm^2) and the 'sampling frequency'. The latter quantity is related to one of the cyclic processes in the brain and is about 20 Hz. Finally, the quantum efficiency of the eye is of fundamental importance. It is comparable to that of photographic emulsions and is of the order of a few per cent.

Studies of the perception of intensity variations reveal many specific details about our visual system. We conclude this section by mentioning one quite illustrative effect, the so-called Mach bands.[2] These bands are seen in Figure 1.19 near the vertical partition lines: to the left of a line, a contrasting bright band or stripe is seen; to the right, a dark band. The effect is an illusion, as clearly demonstrated by covering a neighbouring field to one of the stripes with a piece of paper. The phenomenon derives from the eye's sensitivity to spatial frequencies, and the eye acts in this context as a *differentiator* or *highpass filter* (cf. Section 7.2).

1.4.2 The CCD camera

In a CCD camera, the light-sensitive 'film' is an *integrated light detector* (a CCD chip: charge coupled device), that is, a microelectronic unit based on semiconductor technology. This chip may be regarded as a large number of individual light detectors arranged like the squares on a chessboard. The individual detectors are called the pixels of the CCD (cf. p. 10), and a typical number might be 1000 × 1000 – a spatial resolution comparable to, or exceeding, television quality.

The principle behind the light detection is the fact that when a photon hits, and is absorbed by, the semiconductor silicon, an electric charge in the form of an electron is liberated (more precisely, an electron/hole pair is produced). The charge may be collected and fixed by means of electrodes in the immediate vicinity of the absorbing substrate (Figure 1.20). In practice, a single pixel is attached to several electrodes. In the illustration, of the three electrodes shown, the one in the middle is kept at a higher potential than its neighbours, so that the electrons created are collected here.

The characteristic property of the CCD detector is its *readout*. The readout is accomplished by moving the charges from one electrode to its neighbour; hence,

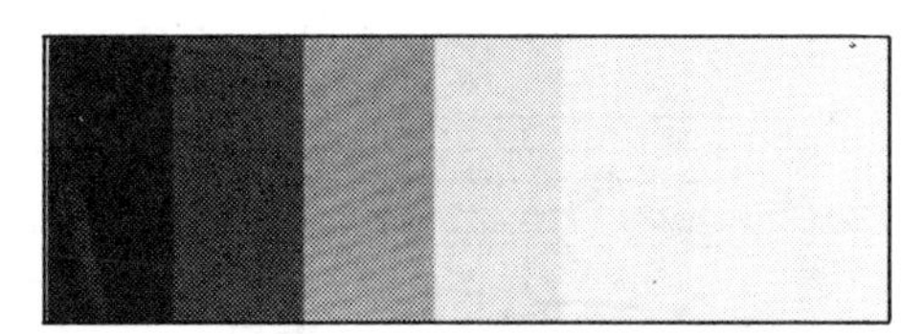

Figure 1.19 Mach bands.

[2] Ernst Mach, Austrian physicist and philosopher (1838–1916)

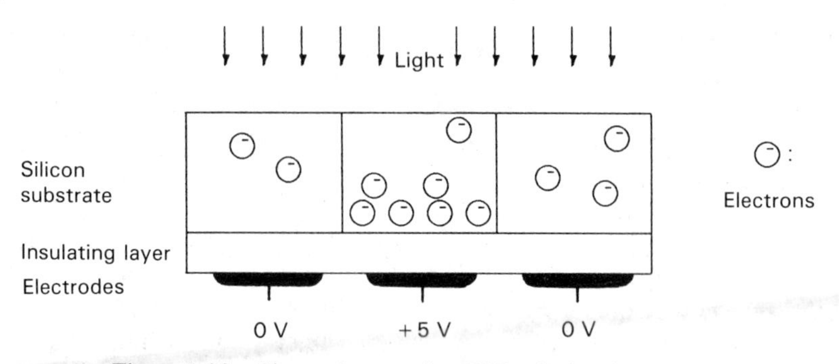

Figure 1.20 The make-up of a CCD pixel (schematic).

after three displacements, the charge has shifted one pixel (Figure 1.21). When the positive voltage is applied to the next electrode in the row, the charges collected follow and, consequently, are forced along the one-dimensional array of pixels shown in the figure. The final recording of the charge takes place after its arrival at the last pixel in the row.

A two-dimensional CCD detector is constructed from a number of one-dimensional detectors like the one described, placed alongside each other and served by the same readout voltages, such as those indicated in Figure 1.21. The charges located in a given row *at right angles* to the direction of transport will thus be obliged to move like a marching column and arrive simultaneously at their respective readout pixels. These readout pixels are, however, built together into another one-dimensional CCD taking over the task of transportation, so that the final readout of the charges takes place from one corner of the two-dimensional array (Figure 1.22).

The quantum efficiency for a typical CCD chip is shown in Figure 1.23 below; as will be seen, the recording of the incident photons is quite complete. While the

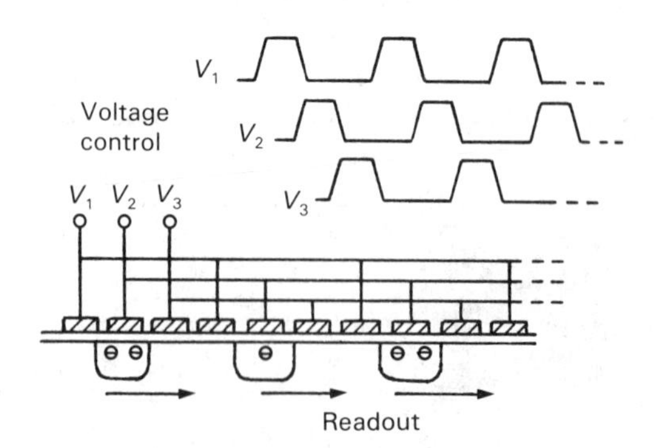

Figure 1.21 A CCD detector and readout voltages.

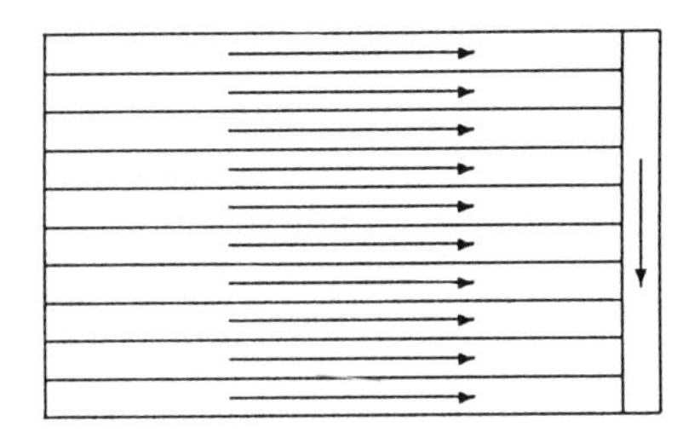

Figure 1.22 Readout across a two-dimensional CCD chip.

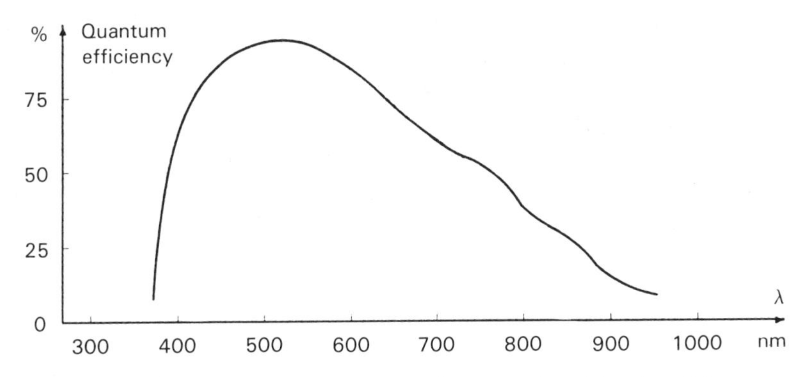

Figure 1.23 The quantum efficiency for a certain CCD chip (RCA).

lack of sensitivity near the red (long-wavelength) end of the curve is caused by the detection principle itself, the problems at the blue end to the left are of a more technical nature and will presumably be solved as CCD technology develops further. Even if the CCD chip is endowed with many excellent properties, certain drawbacks persist, a few of which will be mentioned briefly.

First, the electrons released in the detector substrate are not exclusively triggered by photons – random thermal signals may be responsible as well. Accordingly, one may have to cool the chip in order to suppress the so-called *dark current*. Next, it is understandable that the relatively complicated readout process will introduce erroneous signals – extra or missing electrons – called *readout noise*. Finally, it has proved practically impossible to manufacture the pixels identically; accordingly, *sensitivity variations* across the detector area are to be expected. These differences may be partially eliminated by recording uniformly illuminated surfaces, whereby the chip may be mapped and the sensitivity variations corrected for, either by built-in electronics or by manipulations of the electron numbers recorded.

EXAMPLE 1.11
A CCD chip consists of 1280×640 pixels, as shown below.

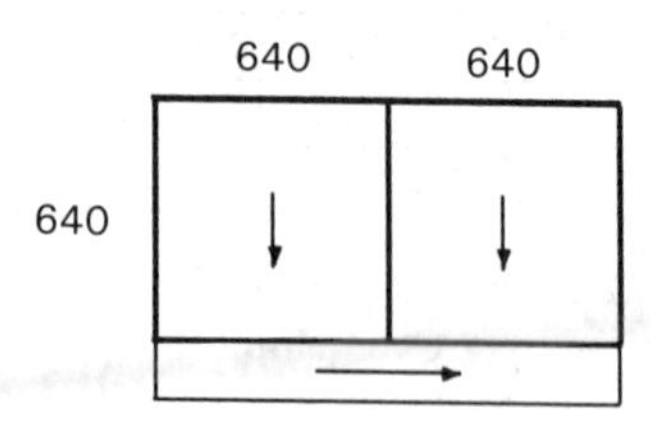

Its two halves are exposed independently to monochromatic light. The light incident upon the left-hand half has a wavelength of 1000 nm; that on the right half, 600 nm. The intensity is, however, in both cases equal to 1.25×10^{-7} W m^{-2}. The quantum efficiencies at the two wavelengths are specified to be 0.1 and 0.6, respectively, and the physical dimensions of the chip are 32×16 mm. The readout takes place from above towards the lower edge, where a one-dimensional CCD collects the charges transferred and transports them from left to right on their way to the final readout pixel.

The amount of charge no. 812 563 consists of 2200 electrons. Find the electron number in portion no. 138 314. Find also the duration of the exposure. The statistical variations in electron number (cf. the next section) should be ignored.

1.5 Detection statistics and the Poisson distribution

As will be clear from Figures 1.2 and 1.3, light detection possesses a markedly statistical character. This character may be summed up in saying that the *light quanta arrive randomly at the detector surface*. For a given time interval, there exist certain probabilities $p_0, p_1, p_2, \ldots$ for the detection of $0, 1, 2, \ldots$ photons, respectively. These probabilities – to be determined presently – will, of course, depend upon the time interval in question. Let us assume that the light intensity is constant in time, so that the relevant probability will depend only on the *elapsed* time – and not on the time *when* the exposure took place.

We start by determining the probability $p_0(T)$ that *no* photons are detected during the interval T. The probability that no photons are detected even during N such intervals, i.e. $p_0(NT)$, may be written as

$$p_0(NT) = p_0(T)^N \tag{1.15}$$

On the other hand, T may be subdivided into M equal portions T/M, so that

$$p_0(T) = p_0(T/M)^M \tag{1.16}$$

A comparison of equations (1.15) and (1.16) shows that for all (rational) numbers $t = N/M$:

$$p_0(tT) = p_0(T)^t$$

where p_0 is completely fixed by choosing $T = 1$:

$$p_0(t) = A^t \tag{1.17}$$

The constant $A = p_0(1)$ denotes the probability that no photons are detected during unit time. The quantity $1 - A$ is thus the detection probability per unit time: the probability that one or more photons are detected.

A more commonly used concept is the so-called *differential detection probability per unit time*. This quantity, here denoted β, appears in the limiting case where t is small (as, for example, when measuring speed). In this limit, the detection probability is

$$\begin{aligned} 1 - p_0(t) &= 1 - e^{t \ln A} \\ &\approx 1 - (1 + t \ln A) \\ &= -t \ln A \end{aligned}$$

so that the detection probability per (small) unit time is

$$\beta = -\ln A \tag{1.18}$$

and finally

$$p_0(t) = e^{-\beta t} \tag{1.19}$$

EXAMPLE 1.12

Explain how the detection probability per unit time is associated with the detection of exactly one photon.

EXAMPLE 1.13

For a certain level of illumination incident upon a CCD pixel, there is a probability of 0.35 that one or more photons will be detected during one second. Find the differential detection probability per second as well as the probability that no photons are detected during

(a) 0.25 s (b) 2 s (c) 10 s.

The remaining probabilities $p_1(t), p_2(t), \ldots$ for detection of exactly $1, 2, \ldots$ photons, respectively, are next determined successively. In Figure 1.24, n photons arriving between times 0 and t are detected first, then one additional photon in the

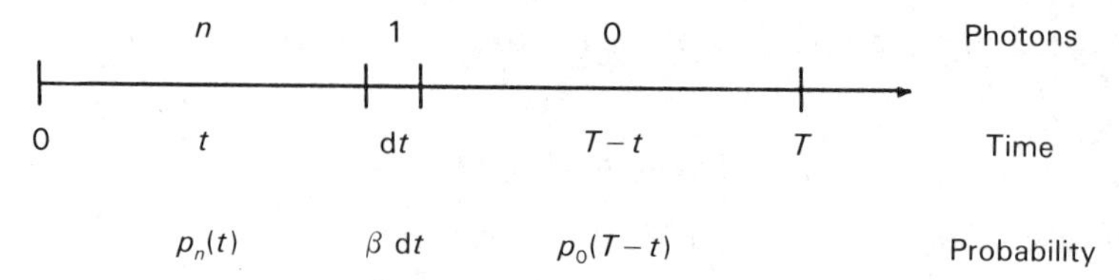

Figure 1.24 Detection of $n+1$ photons.

time interval dt and, finally, zero photons in the remaining interval $T-t$. On multiplication and integration of the probabilities, one finds,

$$p_{n+1}(T) = \int_0^T p_n(t)p_0(T-t)\beta \, dt = \int_0^T p_n(t)e^{-\beta(T-t)}\beta \, dt$$

or

$$e^{\beta T}p_{n+1}(T) = \beta \int_0^T e^{\beta t}p_n(t)$$

This equation states that the quantities $e^{\beta t}p_n(t)$ are obtained from each other by successive integration and multiplication by β, and as $e^{\beta t}p_0(t) = 1$, one obtains

$$p_n(t) = e^{-\beta t}\frac{(\beta t)^n}{n!} \tag{1.20}$$

The statistical distribution

$$p_n = e^{-b}\frac{b^n}{n!} \tag{1.21}$$

is called the *Poisson distribution*[3] *with parameter* b, cf. Appendix A. The result obtained may thus be stated in saying that the *number of detections is Poisson distributed with parameter* $b = \beta t$.

The Poisson distribution has expectation

$$\begin{aligned} E(n) &= \sum_{n=1}^{\infty} np_n = e^{-b}b\sum_{n=1}^{\infty}\frac{b^{n-1}}{(n-1)!} \\ &= b \end{aligned} \tag{1.22}$$

Correspondingly, its second factorial moment is given by

$$E(n(n-1)) = \sum_{n=2}^{\infty} n(n-1)p_n = e^{-b}b^2\sum_{n=2}^{\infty}\frac{b^{n-2}}{(n-2)!} = b^2$$

[3] S. D. Poisson, French mathematician (1781–1840)

so that the variance of the Poisson distribution is

$$V(n) = E(n(n-1)) + E(n) - E(n)^2 = b^2 + b - b^2$$
$$= b \tag{1.23}$$

The distribution parameter b is thus equal to both the expectation (also called the mean) and the variance of the Poisson distribution. In Figure 1.25 the Poisson distribution is depicted for four different parameter values. The choice of the symbol b for the parameter of the Poisson distribution is not a coincidence. We have chosen b to describe the intensity variations (cf. p. 20) across the detector surface, and we conclude this chapter with a brief discussion of this.

From the above derivation it will be seen that the emergence of a Poisson distribution is independent of the duration of the detection. The same holds as regards the detector area (aperture), and the parameter appearing in the resulting distribution must therefore be proportional to both exposure time and detector area. The constant of proportionality β is, in this case, the *differential detection probability per unit time and unit area.*

If the Poisson distribution is specified by a large parameter value, it will appear 'concentrated' around this number (cf. Figure 1.25). This follows from the fact that the standard deviation, i.e. the square root of the variance, is also equal to the square root of the mean value. For large numbers one thus 'always' detects the distribution mean. For smaller detection numbers, the number actually observed – corresponding to a given duration and detector area – is only an 'estimate' (in the statistical sense) of the mean. An improved estimate is only possible through an

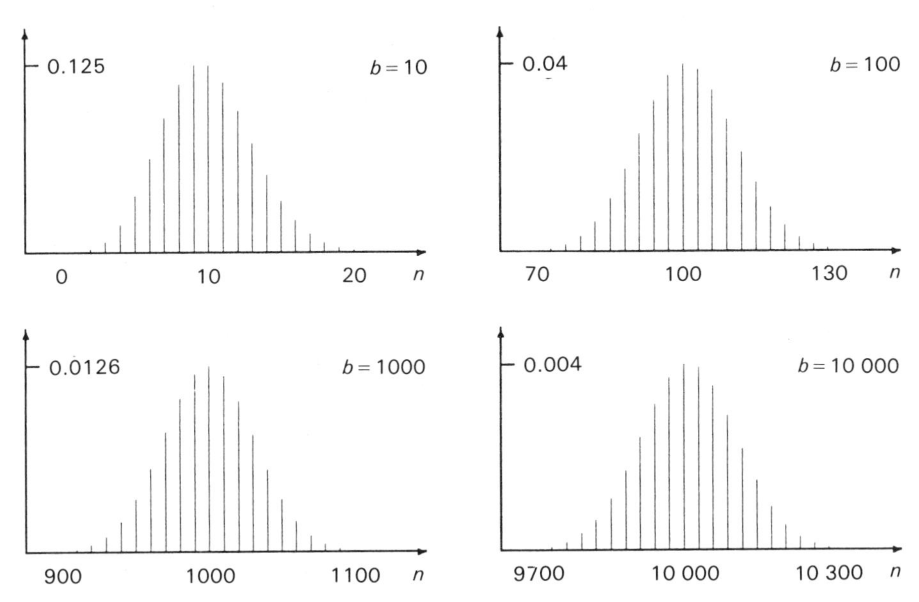

Figure 1.25 The Poisson distribution.

increased photon number – that is, a longer exposure or a larger light-collecting area.

Light intensities must necessarily be defined for large photon numbers. We remind the reader that the detected fraction is termed the quantum efficiency. If this entity is denoted by k and the intensity by I (measured as photon number per unit time and unit area), whereas exposure duration and area are called t and D, respectively, then this limit corresponds to the relation

$$\text{distribution parameter} = \text{number of photons detected}$$

or

$$b = kItD$$

But, according to the derivation above (cf. equation (1.20)), one also has

$$b = \beta tD$$

The differential detection probability per unit time and unit area is thus given by

$$\beta = kI \tag{1.24}$$

expressing the final relation between light intensity and detection number.

In conclusion, we offer some supplementary remarks to the definition of a digital image (see p. 20). The image value $b(x, y)$ of b at the point (x, y) should, strictly speaking, be viewed as an *observation* of a stochastic variable $\mathsf{b} = \mathsf{b}(x, y)$ associated with this point. For those favouring mathematical rigour, a digital image is, therefore, an array of stochastic variables – a so-called *stochastic process* or a stochastic *field*. The 'ideal' digital image – that is to say, that resulting in the limit of very high intensities – thus consists of all the mean values of these variables:

$$\bar{b}(x, y) = E(\mathsf{b}(x, y))$$

and the actual observation $b(x, y)$ at the point (x, y) may be regarded as a statistical *estimate* of the expectation value under consideration.

It may be very helpful to be aware of this rigorous definition of a digital image, although there may seem to be only a remote resemblance to the colour pictures appearing on one's TV screen or computer monitor. For virtually all practical purposes, a digital image will be a rectangular array listing the intensity values appearing in a physical image.

With this knowledge to hand, we are equipped to proceed to investigate the significance of these numbers.

2

Images and Signals

2.1 The concept of information

Even if the commonly used term *information* has rather different meanings depending on the context, no one will disagree that information may be *represented* and *transmitted* in some physical form.

One of the most efficient media for doing so is *images*. The difference between images and other types of *signals* – the quantities representing and transferring the information – is mainly the fact that an image constitutes a surface, a two-dimensional space, over which the signal values are allowed to vary. In contrast, our perception of e.g. sound is one-dimensional, as shown in Figure 2.1. Also, the electric field (cf. Figure 1.11) is a temporal and, consequently, a one-dimensional signal, whereas the electric field in Figure 1.10 is an example of a *spatial* one-dimensional signal. The complete variation of the field given in equation (1.9) does, however, conform to the description of a two-dimensional signal – one that in principle might be regarded as an image.

EXAMPLE 2.1

When viewing a movie, the audience receive a three-dimensional signal $b(x, y, t)$ in the form of an image changing with time. The entire world may, as it were, be described as a four-dimensional signal (one temporal and three spatial dimensions).

The generally enormous amount of information stored in an image is illustrated by the simple example in Figure 2.2a, with the two-dimensional plot in Figure 2.2b. Our ability to exploit the information in this panoramic form rests with the astonishing properties of the human visual system in its interpretation of the signals from the individual 'pixels' in the retina of the eye.

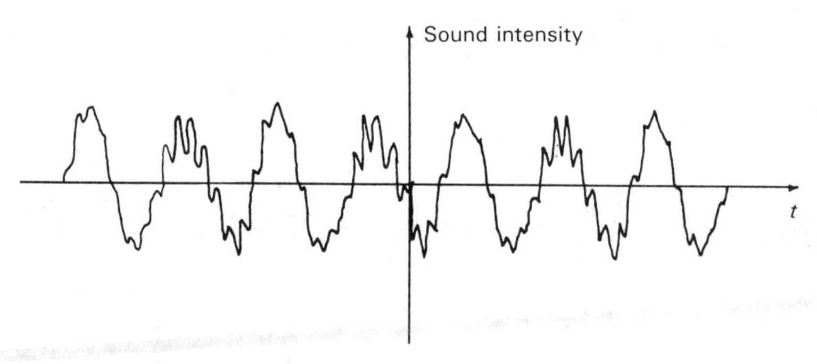

Figure 2.1 Sound curve.

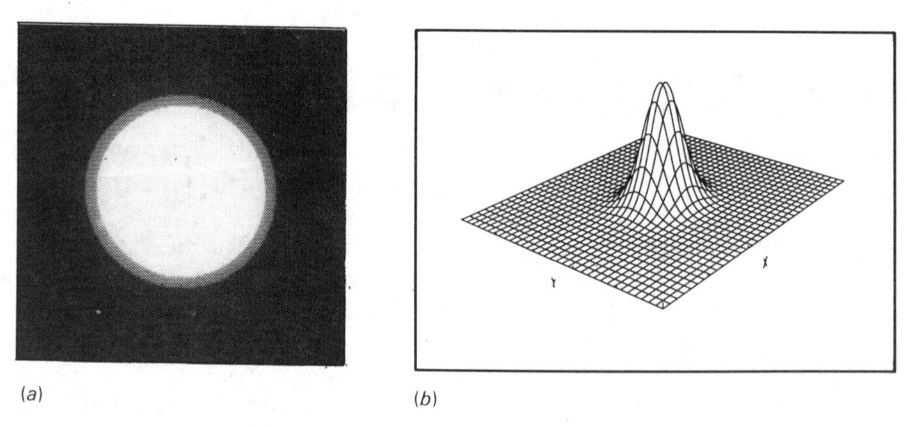

Figure 2.2 Image with a circular object.

The concept of information to be defined and employed in this book facilitates measurements of, and comparisons between, the information content in, for example, the above illustrations. As will be seen, images may well contain vast amounts of information which, however, quite often may be reduced considerably – a fact of the utmost importance in digital image processing.

There is a close connection between the quantitative definition of information and the intuitive one. We are reasonably well equipped to remember and recognize information presented in image form. In fact, we process incredibly large amounts of information in this way – much more than we do verbally.

EXAMPLE 2.2

It seems evident – and may be proved mathematically – that the information content in Van Gogh's *Still Life* (Figure 2.3a) is less than that among the *Fallen Angels* of Brueghel (Figure 2.3b).

Already in Example 1.3 (p. 7) we encountered the problem concerning reduction of image information. We saw that if the partition of an image in $M \times N$ digitization fields – that is, M vertical 'columns' and N horizontal 'rows' – resulted in a satisfactory rendition, then the image information was adequately represented by the MN quantization values.

This number is, in general, 'large' – if television quality is required, for instance, it will be of the order of 1 million. Normally, however, a drastically reduced number will suffice. The condition is that the image intensity, as specified by the quantization value, is approximately constant across extended image areas which may be characterized in a simple manner.

EXAMPLE 2.3

The image in Figure 2.4 consists of only two grey levels, g_0 and g_1. With a choice of 50×30 digitization (pixel size 1×1 mm), one obtains a digital image with 1500 numbers. Where in this image was the following segment taken?

g_0	g_0	g_1	g_1	g_1	g_1	g_1	g_0	g_0	g_0
g_0	g_0	g_1	g_1	g_1	g_1	g_1	g_0	g_0	g_1

If we instead use the code 0 for g_0 and 1 for g_1, the segment is transformed into

0	0	1	1	1	1	1	0	0	0
0	0	1	1	1	1	1	0	0	1

(*a*)

(*b*)

Figure 2.3 Two classical masterpieces.

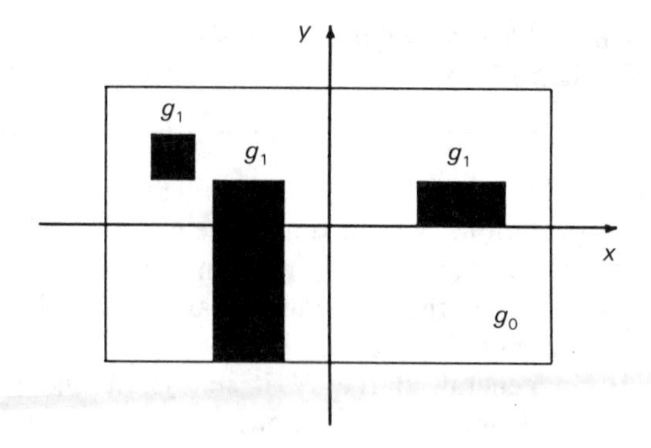

Figure 2.4 Image with two levels.

The total image now consists of 1500 bits. The image is, however, completely specified by the coordinates (k, l) for the lower left-hand corner in each g_1-rectangle, and the dimensions of the rectangle in question, say $K \times L$. The image may now be represented by the following table:

k	l	K	L
10	0	10	5
−20	5	5	5
−13	−15	8	20

This bears a striking resemblance to a digital image! This specification does, however, assume the knowledge of a previously agreed significance of the numbers appearing and is an example of a *coded* image. Here, not only the quantization levels have been coded.

EXAMPLE 2.4

If an image consists exclusively of 'horizontal stripes', that is, if the M quantization levels are one and the same number within each of the N rows (Figure 2.5), the image information may be represented by these N values. The same holds, of course, for images composed of vertical stripes, that is, if each of the M columns consists of N identical values.

One last but important case where the image information may be reduced considerably derives from Example 2.4: Figure 2.5 may be regarded as consisting of *identical columns*. If the image is called $b(x, y)$, one simply has

$$b(x, y) = g(y)$$

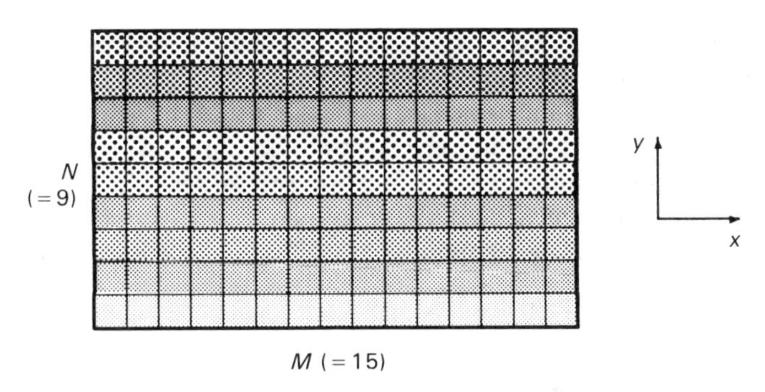

Figure 2.5 Image consisting of horizontal stripes.

If this property is modified slightly, as we now demand identical columns *except for a factor f* – which will then vary with horizontal position x – the image may be written

$$b(x, y) = f(x)g(y) \tag{2.1}$$

An image possessing this property is called *separable*. Such an image consists, then, of rows (or columns) which are identical except for a varying factor.

The image information is now, by virtue of equation (2.1), reduced from the original MN values to the M values for f and the N ones for g – a total of only $M + N$ values; the two-dimensional signal has thus been replaced by two one-dimensional signals. This principle will often be utilized in the following.

EXAMPLE

Among the images below, which ones are separable?

52	36	60
65	45	75
130	90	150
156	108	180
117	81	135
91	63	105
65	45	75

(*a*)

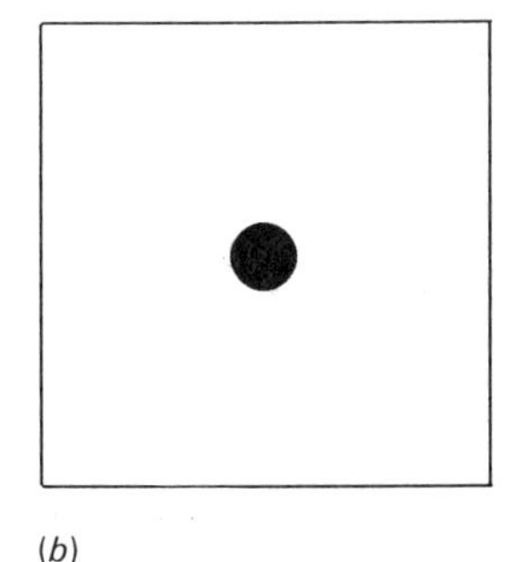

(*b*)

$b(x, y) = e^{-(x^2+y^2)}$

(*c*)

Figure 2.6 Separable images?

Many of the problems dealt with in the first part of this book are related to 'information reduction': How does one represent large amounts of information by means of small amounts, and what are the costs in doing so? This discussion is especially relevant when transmitting information through systems introducing *noise*, that is, uncontrollable signals distorting the information represented. One often imposes the requirement that the information losses due to reduction and noise, respectively, should be comparable.

2.2 Signals

The transfer of information is characterized by a certain degree of 'surprise'. If nothing exciting happens, no change, as seen from the receiver's point of view, no information has been conveyed. The small amount of information in the image Figure 2.4 is closely related to the fact that changes of intensity take place relatively seldom: only along the boundaries of the black rectangles. Normally, changes are present everywhere in an image.

EXAMPLE 2.6

Common speech, as used for communication between humans, is based upon an alphabet, a set of symbols, which may be composed into language. The transfer of information via this medium is, fundamentally speaking, effected in the various sequences of letters being presented to us. The sequence

AAAAAAAAAA

contains no information (except possibly for that appearing when passing from no signal to the first A and/or from the last A to no signal). The sequence

AAAAAAARGH

evidently contains some more.... Here, it is important to distinguish between the possible information content which *might* have been represented by the ten symbols (coded information) and the 'usual' type (verbal information). The sequence

EDKKJFFLJN

might contain coded information, but to make use of it, we need to know the coding principles. The sequence

TO BE OR

might, of course, be a coded message as well, although it obviously contains some verbal information. It is worth mentioning that the verbal information content seems to be almost identical to that conveyed in the longer sequence

TO BE OR NOT TO BE

The latter sequence thus contains superfluous or *redundant* information. The redundancy is of great importance in those cases where noise is present or transmission incomplete. We shall, however, not pursue this topic further in this book.

EXAMPLE 2.7
Information may be transmitted by means of an 'amplitude modulated wave' $f(t) = A(t)e^{i\omega t}$ (Figure 2.7a), as the amplitude $A(t)$ changes with time. In contrast, if one attempts to extract information from the 'pure' harmonic oscillation in Figure 2.7b by measuring the amplitude at different times, one always finds the value A.

EXAMPLE 2.8
An image consisting of a uniform surface contains no information (Figure 2.8a). The image in Figure 2.8b might contain information, which, however, is not immediately accessible. Finally, the image in Figure 2.8c might also contain this type of information – as the intensity is obviously non-constant – but it clearly carries image information.

EXAMPLE 2.9
If Figure 2.8c is one of the items in a slide show, no additional information is obtained (no temporal changes). If, however, it is part of a movie, information is conveyed both via the spatial and temporal changes.

The above quantities, responsible for the transmission of information, will be termed *signals*, as promised. Since the transmission is intimately related to the variation of the signal, this concept is in fact identical to the mathematical counterpart, a *function*. The 'independent' variable is often *time*, as is the case in Figure 2.1 or the sequences in Example 2.6.

In image processing, the simple time variable is replaced by two (spatial) variables corresponding to the two dimensions. An image b, that is, a function $b = b(x, y)$ of two variables, is nevertheless a signal. In what follows, when signals are referred to in general, we shall mainly refer to the independent variable as 'time' – even if the signal in question is an image.

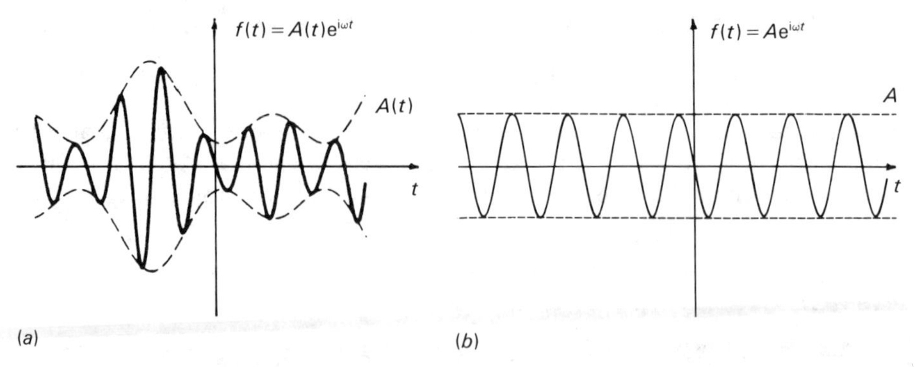

Figure 2.7 Modulated sine wave and its 'carrier wave'.

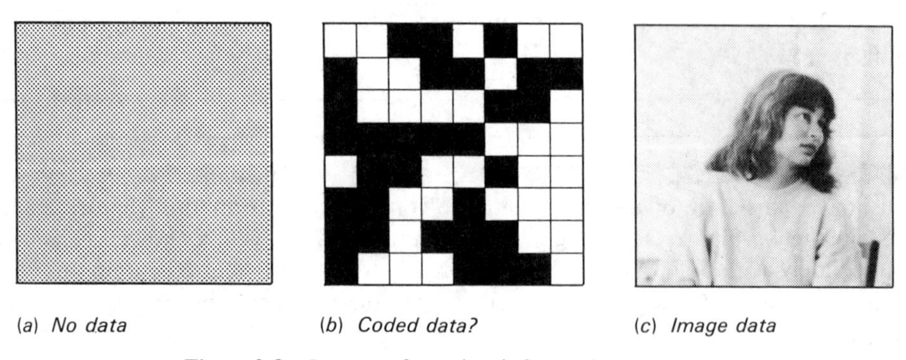

(a) *No data* (b) *Coded data?* (c) *Image data*

Figure 2.8 Images of varying information content.

A signal f, then, assumes certain values $f(t_1), f(t_2), f(t_3), \ldots$ for the values $t_1, t_2, t_3, \ldots$ of time t. If the signal only exists for a finite number N of points in time, it is called *digital*, and the possible values of t are collectively referred to as the *digital time* (more precisely, for the signal in question). On numbering the digital times from 1 to N, one ensures that each signal value will be associated with a unique number between 1 and N. Accordingly, the signal values may be denoted $f[1], f[2], \ldots, f[N]$, and the digital time is represented by the integers from 1 to N.

Even if infinitely many signal values $f[1], f[2], \ldots$ or $\ldots, f[-2], f[-1], f[0], f[1], f[2], \ldots$ are present, the signal will still be termed digital. If, on the other hand, the variable t is *continuous*, that is to say, it assumes values in intervals of real numbers, then the signal too is referred to as being continuous, or an *analog signal*. The mathematical synonym is a 'continuous function', whereas a digital signal would normally be called a vector.

We shall use the notation

$$f(t), \qquad t \in (t_1, t_2)$$

for an analog signal f defined for values of t in the open interval from t_1 to t_2. In the same way, we shall write

$$f[n], \qquad n \in [n_1, n_2]$$

for the digital signal f defined for $n = n_1, n_1 + 1, \ldots, n_2 - 1, n_2$, that is, the integers in the closed interval from n_1 to n_2. Another possible notation is $f = (f[n_1], f[n_1 + 1], \ldots, f[n_2])$.

Two signals f and g are identical if and only if they consist of the same values:

$$f = g \Leftrightarrow (f[n_1], \ldots, f[n_2]) = (g[n_1], \ldots, g[n_2])$$

The term 'analog' for signals covers, in fact, a limiting situation where the digital times are so narrowly spaced that the corresponding signal values do not change considerably – in a context-sensitive way, for example, as regards information content – when going from one digital time to the next one (Figure 2.9). What is the difference, then, between an analog signal and a digital one? A 'genuine' analog signal $f(t)$ may immediately be converted into a digital signal by selecting times $t_1, t_2, t_3, \ldots$ and choosing

$$f[n] = f(t_n) \tag{2.2}$$

This procedure, where the digital signal is produced pointwise from the analog one,

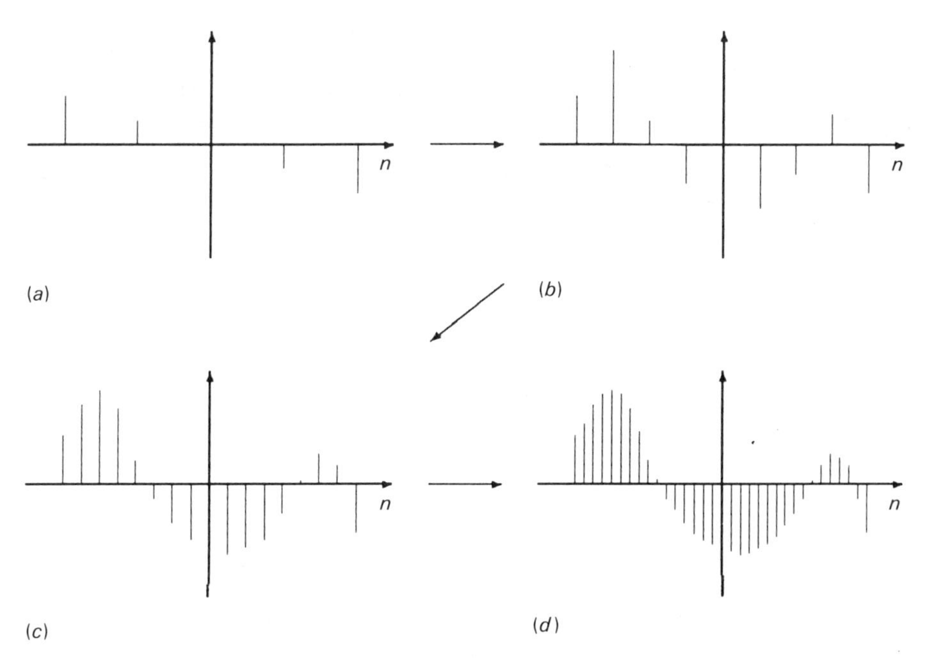

Figure 2.9 Transition from digital signal to analog signal.

is called *sampling* (cf. Figure 2.10). Due to the various choices of digital time, the conversion is not unique.

Nor is the inverse problem – that of reconstructing an analog signal from a digital one consisting of its samples – solvable in a unique way, as illustrated in Figure 2.11. But, as indicated by Figure 2.9, the reconstruction intuitively becomes progressively more unique, the denser the grid of digital times.

Later (p. 67) we shall discuss the so-called *sampling theorem*, according to

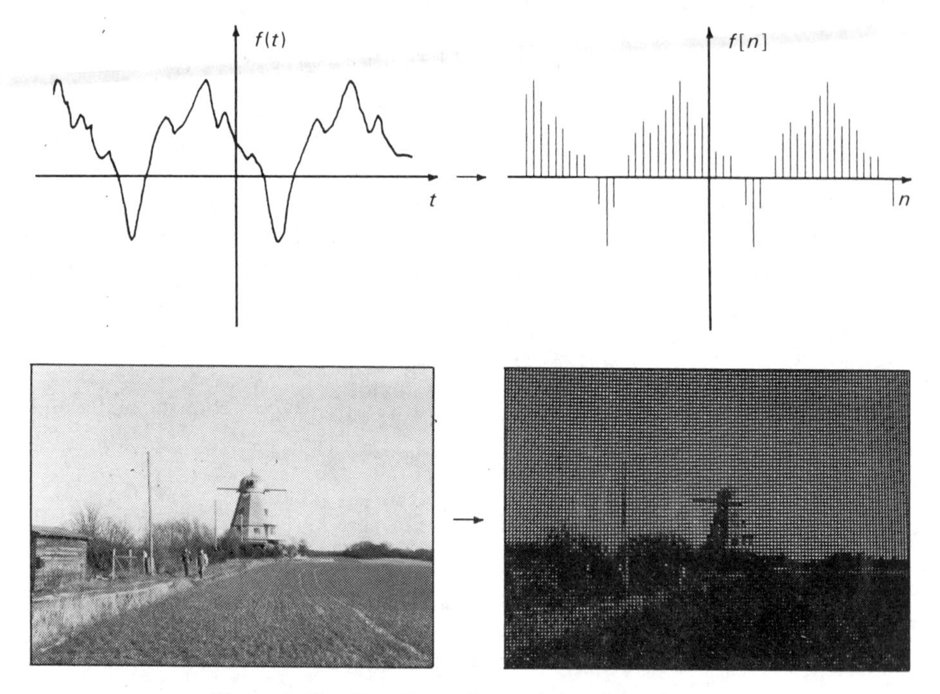

Figure 2.10 Sampling of sound signal and image.

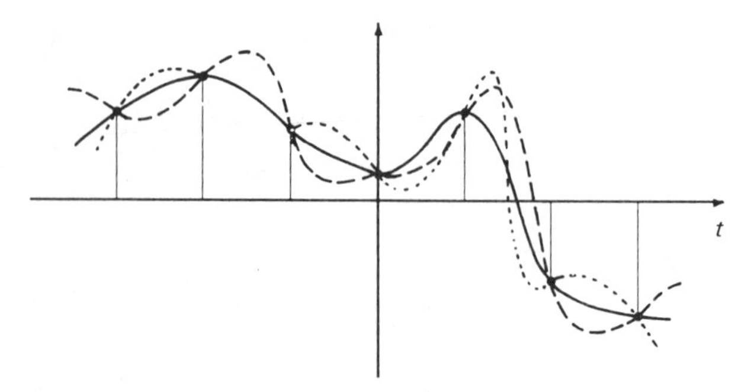

Figure 2.11 Different analog signals having the same samples.

which an analog signal may be reconstructed uniquely from its samples, provided it does not contain components of higher frequency than one-half of the sampling frequency.

EXAMPLE 2.10

Digital sound, as recorded on a compact disc, is normally sampled at a frequency of 44 100 Hz. The maximal frequency in the reconstructed sound is, therefore, $44\,100/2 \simeq 22$ kHz.

EXAMPLE 2.11

If a draftsman is asked to join the points in Figure 2.12a by a 'smooth' curve, he will almost certainly – the vague formulation notwithstanding – regard the task as well defined. Figure 2.12b shows a professional solution.

EXAMPLE 2.12

We regard the temperature in a room as a continuous function of position within the room – that is, as an analog signal. Nevertheless, it is normally quite enough to have one thermometer in the room, as the temperature variations are insignificant. The lack of knowledge ('undersampling') of the precise temperature at various points in the room is thus counterbalanced by the advantages of the simple, digital representation of the (in principle) enormously large number of temperature values.

As illustrated by Example 2.12, a digital representation of an analog signal may serve the purpose of information reduction – *not* conservation of the complete information content present in the signal, as attempted in Example 2.11.

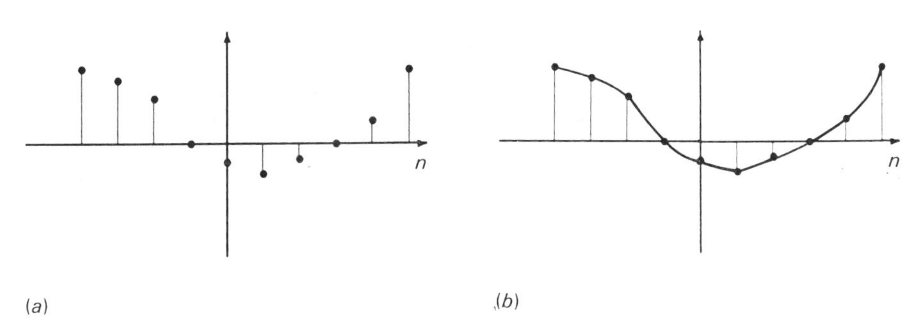

Figure 2.12 A drawing problem and its solution.

2.3 Digital representation of analog signals

An analog signal $f(t)$ might, in principle, exist for all values of time t, both past and future. In practice, however, we are restricted to processing only the values occurring in a finite interval of time, and the signal f must then be *truncated* as in Figure 2.13. Here, the information originating outside the truncation interval is lost. In a number of cases, however, no losses occur. The two most important ones are the following.

1. The signal f is *periodic*. This means that there exists a number T such that

$$f(t+T) = f(t) \qquad \text{for all } T \tag{2.3}$$

Whatever the instantaneous signal value, it reappears after T seconds, as illustrated in Figure 2.14. The shape of the signal over any time interval of length T is thus repeated after T seconds. In place of T in equation (2.3), we might instead have used $2T, 3T, \ldots$. The smallest possible value for T in equation (2.3) is called the *fundamental period* of the signal (or simply its period).

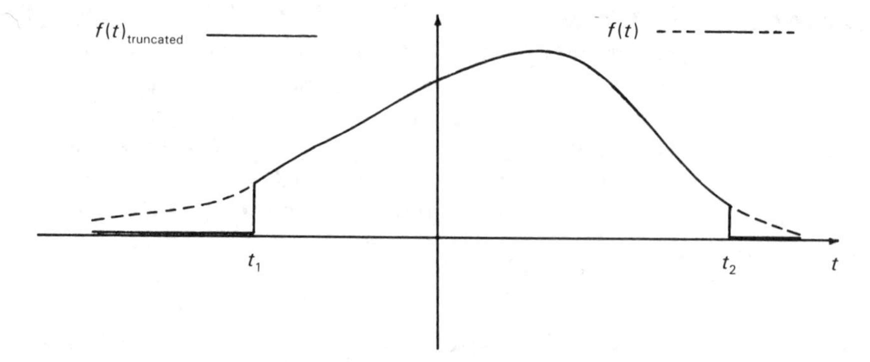

Figure 2.13 A signal $f(t)$, truncated in the interval from t_1 to t_2.

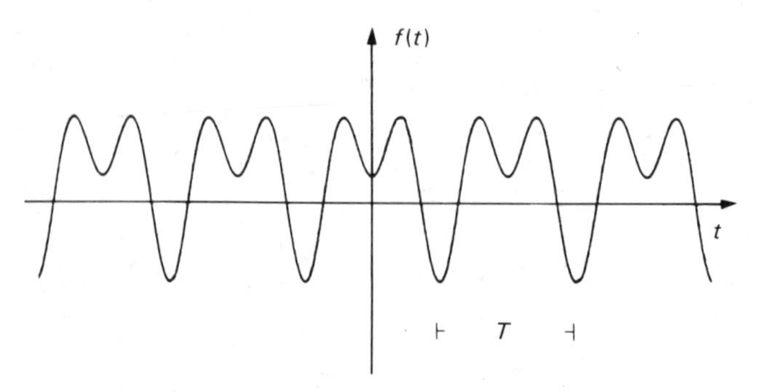

Figure 2.14 A periodic signal $f(t)$ with period T.

Among the periodic signals, the most important are the harmonic ones:

$$f(t) = Ae^{i\omega t}$$

where the amplitude A is allowed to be complex. However, the *angular frequency*

$$\omega = 2\pi/T$$

is a real number.

2. The signal f is *transient*, that is, 0 *ab initio* outside a certain, finite, interval of time. The term is used also if the signal approaches zero for large values of $|t|$, for example, as indicated in Figure 2.13. In this case, too, one may select a finite interval outside which the signal vanishes to a sufficient degree of accuracy. If this approximation is inadequate, *edge effects* are said to be present.

EXAMPLE 2.13

If the note 'middle A' is intonated on a flute, a vibration of frequency 440 Hz and a few of its *harmonics* or *overtones* (oscillations of frequency $2 \times 440 = 880$ Hz, $3 \times 440 = 1320$ Hz, etc.) are generated. After a few seconds, the ear or the microphone has received thousands of repetitions of the fundamental period $T = \frac{1}{440}$ s. For the processing of this signal, it will thus be irrelevant whether its duration is considered finite or infinite.

For electromagnetic oscillations, this consideration may be even more relevant. For laser light, 10^{14}–10^{15} repetitions of the fundamental oscillation is quite normal.

EXAMPLE 2.14

A photograph represents, of course, only part of the scene present before the eyes of the photographer. With a 'good' photograph, however, the action outside the frame is irrelevant in the actual context, and the finite size of the image will not be important. Figure 2.15 shows two examples of truncation.

For the above reasons, we restrict ourselves to analog signals which are only 'active' over a finite interval of time. The information content conveyed by such signals is, in practice, limited by the sensitivity of the receiving system with respect to small variations in signal value – especially so if noise is present.

These limitations imply that signal values differing by less than a certain amount should be considered equal – at least with respect to information content. The limitations may, however, prove to be distinct advantages!

If, that is to say, the quantity ε sets the upper limit to significant variations in signal value, one can subdivide the interval into N equal parts, so as to ensure a signal variation less than ε within each part, as shown in Figure 2.16. The original

Figure 2.15 Mermaid and Marilyn: truncated images.

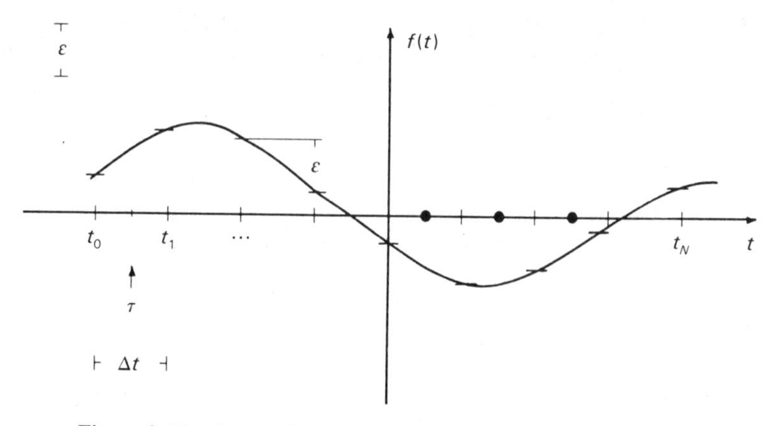

Figure 2.16 An analog signal with 'representative' intervals.

analog signal, with its continuum of values, may accordingly be boiled down to N numbers representing the signal values in the N intervals in a complete fashion.

The most commonly used method for specifying the N numbers uses the values in the mid-points of the subdivision intervals:

$$f(\tau), f(\tau + \Delta t), f(\tau + 2\Delta t), \ldots, f(\tau + (N-1)\ \Delta t)$$

where Δt is the size of the subdivision interval and τ is the first of these sampling mid-points. Obviously, a smaller N may be sufficient if one abandons the requirement of subdivisions of equal size: for instance, the intervals marked by the symbol • in Figure 2.16 might have been pooled.

If the total time interval ranges from 0 to T, then $\Delta t = T/N$, $\tau = \Delta t/2$, and

$$f[n] = f((n + \tfrac{1}{2})T/N), \qquad n \in [0, N-1] \tag{2.4}$$

A notionally simpler digitization consists in choosing

$$f[n] = f(nT/N), \qquad n \in [0, N-1] \tag{2.5}$$

As will be seen, there is normally an optimal value of N where all the relevant information has been represented, and represented only once. Larger values of N will result in *oversampling*, smaller ones in *undersampling*.

No existing physical system is capable of recording an instantaneous signal value. In practice, a measurement always requires the passage of a certain interval of time, and a physically more reasonable digitization scheme would then *average* over the subdivision intervals:

$$f[n] = \frac{1}{\Delta t} \int_{n\,\Delta t}^{(n+1)\,\Delta t} f(t)\,\mathrm{d}t \tag{2.6}$$

Exactly this type of digitization was carried out for the photographs in Chapter 1. The integral sign corresponds to the total number of photographic grains, and on dividing by the interval length, one obtains the grain density – or, rather, its average value.

The digitization in equation (2.6) obeys the rule

$$\sum_{n=0}^{N-1} f[n] = \frac{1}{\Delta t} \int_0^T f(t)\,\mathrm{d}t$$

a relation which, for other types of digitization, only holds in the analog limit. As $\Delta t = T/N$, we have the important result

$$\frac{1}{N} \sum_{n=0}^{N-1} f[n] \to \frac{1}{T} \int_0^T f(t)\,\mathrm{d}t \qquad \text{for } N \to \infty \tag{2.7}$$

The expression which replaces equation (2.6) in the two-dimensional case is

$$b[m,n] = \frac{1}{\Delta x\,\Delta y} \int_{m\,\Delta x}^{(m+1)\,\Delta x} \int_{n\,\Delta y}^{(n+1)\,\Delta y} b(x,y)\,\mathrm{d}y\,\mathrm{d}x \tag{2.8}$$

stating, simply, the definition of the integral in a symmetrical fashion.

EXAMPLE 2.15

A signal f is given by

$$f(t) = 5 \sin t + \sin 2t, \qquad t \in (0, 2\pi)$$

It is being recorded by a signal processing system which, however, is unable to distinguish signal values differing less than 0.9. How many equidistantly spaced points of time in the interval are necessary in order to perform a sampling of f? The possible choices are:

(a) 30 (b) 40 (c) 50 (d) 60.

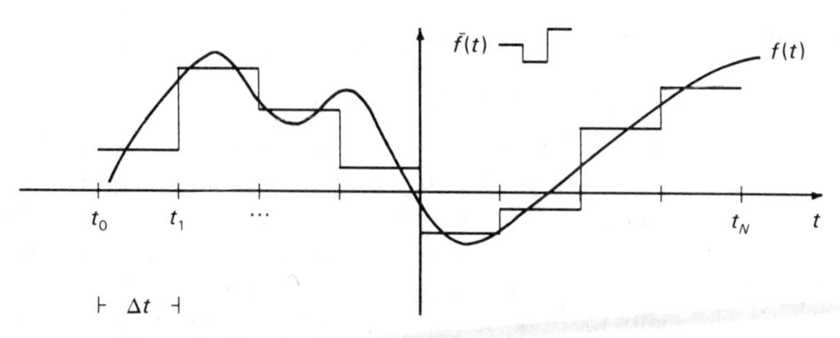

Figure 2.17 An analog signal and its reconstruction.

EXAMPLE 2.16

The signal f in Example 2.15 is digitized by means of each of the three methods from equations (2.4), (2.5), and (2.6), for $N = 40$. Find the values

(a) $f[0]$ (b) $f[1]$ (c) $f[10]$ (d) $f[20]$ (e) $f[39]$.

When reconstructing the image in Figure 1.5, we agreed to distribute the quantization numbers evenly across each of their individual fields. For one-dimensional signals, this procedure is illustrated in Figure 2.17. Note that the reconstruction $\bar{f}$ of f again is an analog signal (even if the discontinuities render the term 'analog' slightly misleading). We shall return later to physically more acceptable reconstruction or *interpolation* methods (Sections 2.7 and 6.5).

In the digitizations of the analog signal f or its analog reconstruction (Figure 2.17), an optimal representation of the available information in f is possible. For various purposes of economy, one will often tend to reduce the information content further. We are thus led to the problem of approximating digital signals with others, the properties of which allow suitable compromises between, on the one hand, 'information conservation' and, on the other hand, reasonable demands on available resources.

The simplest digital and analog signals for this purpose are described in the next section.

2.4 Special signals

In this section, we introduce a few special signals and images which will be used extensively later. We shall assume that the signals in question are defined for all positive and negative values of time (including zero). For images, this means that the extension is, in principle, infinite.

Of fundamental importance is the *delta signal* or δ-signal; this is shown in Figure 2.18a. It is defined as follows:

$$\delta[n] = \begin{cases} 1 & \text{for } n = 0 \\ 0 & \text{for } n \text{ positive or negative} \end{cases}$$

This signal, also commonly referred to as the *unit impulse*, also exists in 'time-translated' versions. For instance, the signal $\delta[n+2]$ consists of zeros, except for the digital time $n = -2$ (Figure 2.18b); here, the signal is 1. Generally,

$$\delta[n-k] = \begin{cases} 1 & \text{for } n = k \\ 0 & \text{otherwise} \end{cases} \tag{2.9}$$

It is *very* important to note the distinction between the entities k and n. Here, k is a certain number, a *parameter*, whereas n is the digital time, that is, a *symbolic variable*, part of the signal notation.

For images, we correspondingly define

$$\delta[m-k, n-l] = \begin{cases} 1 & \text{for } m = k \text{ and } n = l \\ 0 & \text{otherwise} \end{cases}$$
$$= \delta[m-k]\,\delta[n-l]$$

where the two-dimensional delta signal has been expressed as a product of two one-dimensional delta signals. The image version of the delta signal is thus *separable*. In the notation for the image, m and n appear as variables – we might call $[m, n]$ the *digital position*. In contrast, k and l are parameters, and $[k, l]$ specify the 'location' of the δ-signal (Figure 2.19). In order to represent the δ-signal as an image, the rendition here gives the 'reconstructed analog version' (cf. also Figure 2.17). The value 0 corresponds to white, the value 1 to black.

The distinction between symbolic variables and parameters is especially evident in the important relation below. Here, the arbitrary signal f is written as a linear combination of δ-signals:

$$f[n] = \sum_{k=-\infty}^{\infty} f[k]\,\delta[n-k] \tag{2.10}$$

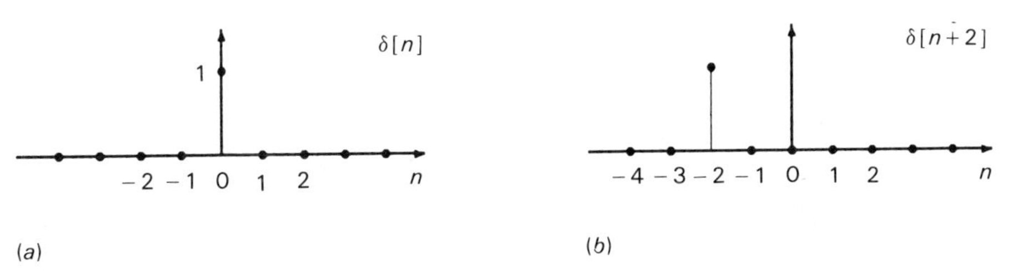

Figure 2.18 The digital delta signal.

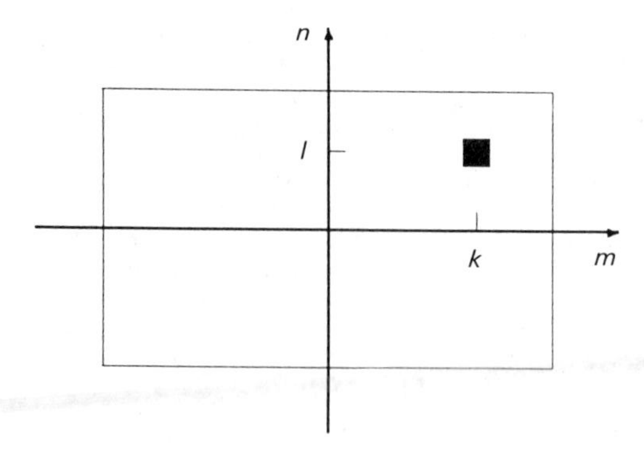

Figure 2.19 The two-dimensional delta signal.

This decomposition has been illustrated pictorially in Figure 2.20, both in the one-dimensional case (top) and for images (bottom).

Another important signal is the *step* signal

$$s[n] = \begin{cases} 1 & \text{for } n \geqslant 0 \\ 0 & \text{otherwise} \end{cases} \tag{2.11}$$

depicted in Figure 2.21. In the analog case, the definition of the step signal is obtained from expression (2.10) on substitution of t for n and parentheses for square

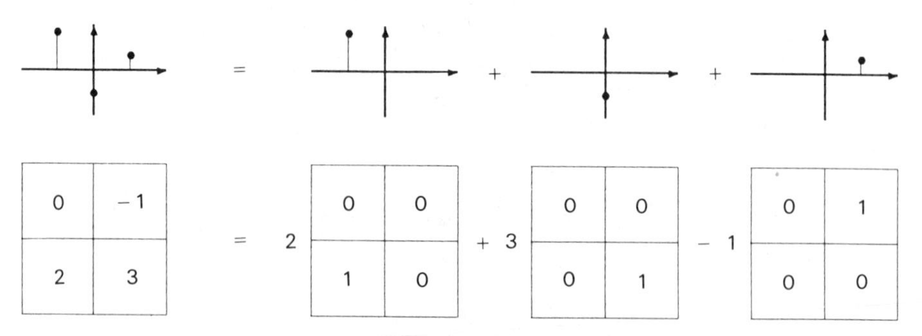

Figure 2.20 Delta-signal decomposition.

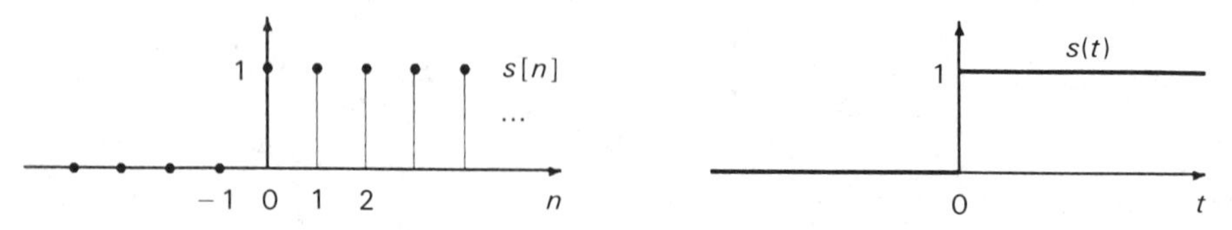

Figure 2.21 The unit step signal.

brackets. In the same way, the two-dimensional step signal is defined by

$$s[m, n] = \begin{cases} 1 & \text{for } m \geqslant 0 \text{ and } n \geqslant 0 \\ 0 & \text{otherwise} \end{cases} \tag{2.12}$$

This image, too, is separable, a product of two one-dimensional step signals:

$$s[m, n] = s[m]\, s[n]$$

The analog image step signal is shown in Figure 2.22.

If the step signal is expressed as a linear combination of unit impulses, the result is

$$s[n] = \sum_{k=0}^{\infty} \delta[n-k]$$

Just as was the case for the delta signal, the step signal exists in displaced variants $s[n-k]$.

EXAMPLE 2.17

Express $3\delta[n-2] - 2\delta[n+1]$ as a linear combination of step signals.

EXAMPLE 2.18

Is the equation $s[k] = \Sigma_{j=-\infty}^{k} \delta[j]$ correct if j and/or k is regarded as (a) parameter(s), (b) digital time?

From the step signal, we can form *rectangular* signals

$$r_N[n] = \begin{cases} 1 & \text{for } n \subset [0, N-1] \\ 0 & \text{otherwise} \end{cases}$$

$$= s[n] - s[n-N]$$

which are zero for all values of n except $0, 1, 2, \ldots, N-1$. The definition of the rectangular signal is clear from Figure 2.23, both in the digital and analog cases.

The definition of the two-dimensional rectangular digital signal is obvious:

$$r_{M,N}[m, n] = r_M[m]\, r_N[n]$$

A similar definition holds in the analog case.

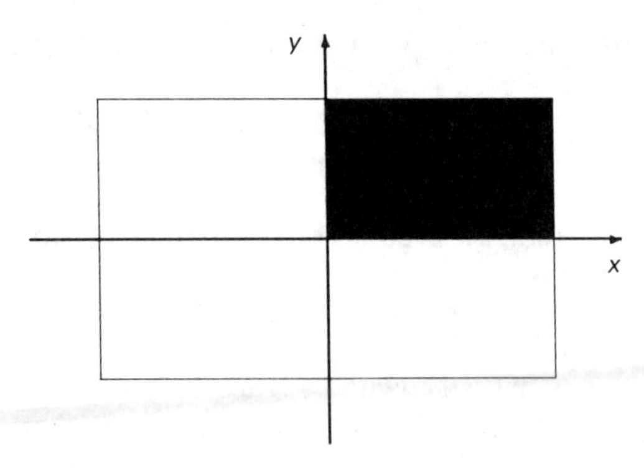

Figure 2.22 The two-dimensional step signal.

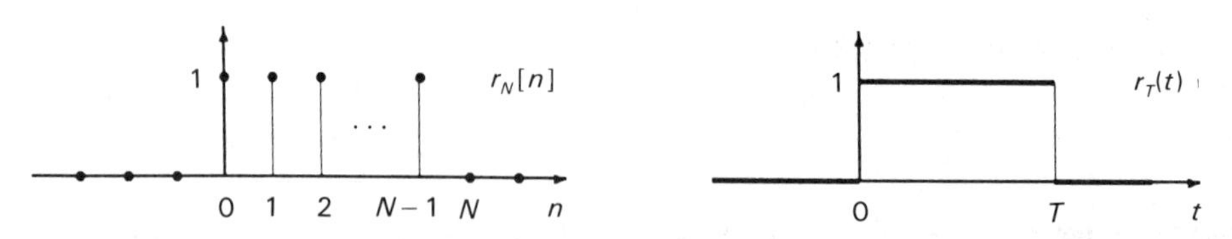

Figure 2.23 The rectangular signal.

EXAMPLE 2.19

Referring to the table on p. 34, the image in Figure 2.4 can be specified as follows:

$$b(x, y) = (g_1 - g_0) \cdot r_5(x + 20) r_5(y - 5) + (g_1 - g_0) \cdot r_8(x + 13) r_{20}(y + 15) \\ + (g_1 - g_0) \cdot r_{10}(x - 10) r_5(y) + g_0$$

More generally, a reconstructed analog signal may be expressed as a linear combination of rectangular signals:

$$\bar{f}(t) = \sum_{k=0}^{N-1} f[k] r_{\Delta t}(t - t_k)$$

where the meaning of the quantities Δt and t_k is clear from Figure 2.17. In the linear combination above, the digitized values $f[k]$ appear as coefficients for the N rectangular signals present.

EXAMPLE 2.20
A 4×3 digital image has been digitized as follows:

143	338	234	65
99	234	162	45
165	390	270	75

The digitization has been carried out by means of square pixels, the size of which is unity. Prove that the image is separable, and express the reconstructed analog image $\bar{b}(x, y)$ in terms of one-dimensional rectangular signals. The coordinate system should be placed in the image centre.

A final but important type of signal is the *exponential* signal (sometimes also referred to as the *quotient* signal):

$$\exp_z[n] = z^n \tag{2.13}$$

where z is an arbitrary, possibly complex, number. This number is called the *base* or *quotient* of the signal.

If z is written in the usual form $z = Ae^{i\theta}$, $\exp_z[n]$ is decomposed into a product of a real exponential A^n and a harmonic signal $e^{i\theta n}$. We shall consider these types one at a time. They are shown below in Figure 2.24, as are their analog versions.

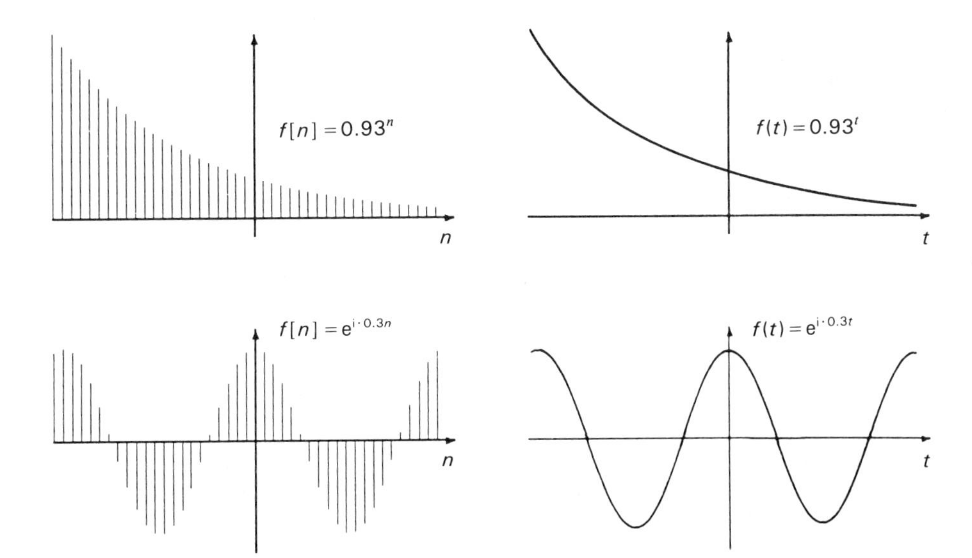

Figure 2.24 Exponential and harmonic signals.

Note that the harmonic analog signal is always periodic. In order to utilize this important property in the digital case it will often be assumed that the base z is one of the *complex unit roots* $w_N = e^{i2\pi/N}$. If so, the digital harmonic signal $f[n] = w_N^n$ consists only of N different values

$$(f[0], f[1], \ldots, f[N-1]) = (1, w_N, w_N^2, \ldots, w_N^{N-1})$$

Lastly, the two-dimensional exponential signal is expressed as follows:

$$\exp_{z,w}[m, n] = z^m w^n \tag{2.14}$$

which is again a separable image.

When employing the simple signals and images described here for approximation purposes, a method for specifying the quality of a signal approximation is called for. In other words, we need to know how to measure the distance between two signals!

2.5 Signal distance

The distance between two points x and y on the real axis may be written in the well-known form $|x - y|$. Thus $|x|$, the absolute value of x, denotes distance between the point x and the origin.

In introducing the distance between the signals f and g, the same notation is used, and we utilize the economy inherent in defining only $|f|$, here called the *norm* of the signal f. This quantity is taken to measure the distance between f and the zero signal (the signal consisting of only zeros), and the distance between f and g is then defined as $|f - g|$, that is, *the norm of the difference signal.*

The distance on the real axis may, if one wishes, be written as the positive square root of x^2:

$$|x| = \sqrt{x^2} \tag{2.15}$$

If now x is a point in the plane, say $x = (x_1, x_2)$, its distance from 0 is

$$|x| = \sqrt{x_1^2 + x_2^2} \tag{2.16}$$

If, finally, x is a point $x = (x_1, x_2, x_3)$ in space, its distance from the origin is

$$|x| = \sqrt{x_1^2 + x_2^2 + x_3^2} \tag{2.17}$$

In equations (2.16) and (2.17), the norm for x is found from the sum of squares of the component distances from 0 – cf. equation (2.15). These relations *define* the quantity $|x|$, here denoting the length of the vector x.

It seems quite natural to generalize the above expressions to vectors having more than three components, that is, digital signals, in spite of the lack of a direct geometric interpretation. In doing so, we cater for possible complex signal components. A slight difference arises from the fact that the distance of a complex number x from zero, that is, its modulus or absolute value $|x|$, is

$$|x| = \sqrt{xx^*}$$

where x^* is the complex conjugate of x.

Thus, for a signal $f[n]$, $n \in [0, N-1]$, possibly complex, the *norm* is defined by

$$|f|^2 = \sum_{n=0}^{N-1} |f[n]|^2 = \sum_{n=0}^{N-1} f[n] f[n]^* \tag{2.18}$$

As all the quantities occurring in the summation are real and positive, the norm $|f|$ is also real and positive. It is zero if and only if f is the zero signal. Consequently, the distance between two signals only vanishes if the two signals are identical.

As in geometry, we finally introduce the *scalar product* $f \cdot g$ of the signals f and g. It is defined (cf. equation (2.18)) as

$$f \cdot g = \sum_{n=0}^{N-1} f[n]\, g[n]^* \tag{2.19}$$

The scalar product is in general complex, and from the definition it is seen that $(f \cdot g)^* = g \cdot f$. The ordering of the factors in the scalar product is thus essential.

EXAMPLE 2.21

The distance between two complex numbers $x = x_1 + \mathrm{i}x_2$ and $y = y_1 + \mathrm{i}y_2$ can be found by regarding x and y as pairs of real numbers and using equation (2.16). Show that the formula (2.18) yields the same result.

EXAMPLE 2.22

Find the norm of the signals f and g, where $f = (3 + \mathrm{i}, 1, -\mathrm{i}, 2\mathrm{i})$ and $g = (0, 1, \mathrm{i}, -\mathrm{i})$. Also, find the distance between them as well as the scalar products $f \cdot g$ and $g \cdot f$.

The above formulae may be generalized, in an obvious way, to the two-dimensional case. For instance, the scalar product of two images b and c is defined as

$$b \cdot c = \sum_{m=0}^{M-1} \sum_{n=0}^{N-1} b[m, n]\, c[m, n]^* \tag{2.20}$$

EXAMPLE 2.23

Find the norm of the image b given by

$$b[m, n] = s[m, n] \exp_{1/2, 1/3}[m, n] = \begin{cases} (1/2)^m (1/3)^n & \text{for } n, m \geqslant 0 \\ 0 & \text{otherwise} \end{cases}$$

Finally, we make use of the above definitions to find the norm, distance and scalar product for analog signals in a finite interval of time – say, for simplicity, the interval from 0 to T. As previously, it will suffice to define the scalar product $f \cdot g$ between the signals f and g. The norm may be introduced as $|f| = \sqrt{f \cdot f}$, and the distance between f and g becomes $|f - g|$.

The analog signals $f(t)$ and $g(t)$ in the interval $[0, T]$ are now regarded as being represented by the digital signals $f[n]$ and $g[n]$, $n = 0, 1, \ldots, N-1$, via the sampling

$$f[n] = f(nT/N) \qquad \text{and} \qquad g[n] = g(nT/N)$$

If so, the scalar product is

$$f \cdot g = \sum_{n=0}^{N-1} f[n]\, g[n]^*$$

For $N \to \infty$, the scalar product may now be approximated by an integral (cf. expression (2.7)), and the transition digital → analog is expressed in the limit

$$\frac{1}{N} \sum_{n=0}^{N-1} f[n]\, g[n]^* \to \frac{1}{T} \int_0^T f(t) g(t)^* \, \mathrm{d}t$$

If, then, the scalar product between two analog signals f and g is introduced as

$$(f \cdot g)_{\text{analog}} = \int_0^T f(t) g(t)^* \, \mathrm{d}t \tag{2.21}$$

one obtains the following convenient relationship, linking the scalar products in the digital and analog cases:

$$\frac{1}{N} (f \cdot g)_{\text{digital}} = \frac{1}{T} (f \cdot g)_{\text{analog}} \tag{2.22}$$

The norm of a real analog signal f is accordingly given by the square root of

$$|f|^2 = \int_0^T |f(t)|^2 \, \mathrm{d}t \tag{2.23}$$

This quantity is often referred to as the (total) *energy* of the signal (cf. Sections 1.2 and 1.3), while $|f|^2/T$ is called the *power* of the signal.

EXAMPLE 2.24

Find the energy and power for the following signals:

(a) $f(t) = 0.93^t$, $t \in (0, 7)$

(b) $f(t) = \cos(0.3t)$, $t \in (0, 7)$

According to equation (2.22), the appropriate expression for the power in a digital $f[n]$, $n \in [0, N-1]$, is

$$\frac{1}{N} |f|^2 = \frac{1}{N} \sum_{n=0}^{N-1} |f[n]|^2 \tag{2.24}$$

EXAMPLE 2.25
Find the power in the digital signals obtained from two analog signals in Example 2.24 by equidistant (mid-point) sampling with

(a) $N=2$ (b) $N=4$ (c) $N=8$.

From the above it follows that, when using different digitizations of the same analog signal, *sums* are conveniently replaced by *averages*.

EXAMPLE 2.26
An analog signal $f(t)$ is of constant value f_0 over a certain interval of time. If f is digitized by sampling, the norm of $f[n]$ is given by

$$|f|^2 = \sum |f[n]|^2 = N f_0^2$$

When using $|f|/\sqrt{N} = f_0$ as the quantity measuring the deviation between f and the zero signal, independence of N is obtained.

The terms energy and power are used for images b as well; they are, respectively, $|b|^2$ and $|b|^2/A$, where

$$|b|^2 = \iint |b(x, y)|^2 \,\mathrm{d}x \,\mathrm{d}y$$

and where the integration is extended over the image area A.

EXAMPLE 2.27
In this example, two images b and c are specified as follows:

$$b(x, y) = xy^2$$
$$c(x, y) = x^2 y$$

within the rectangle shown in the figure below.

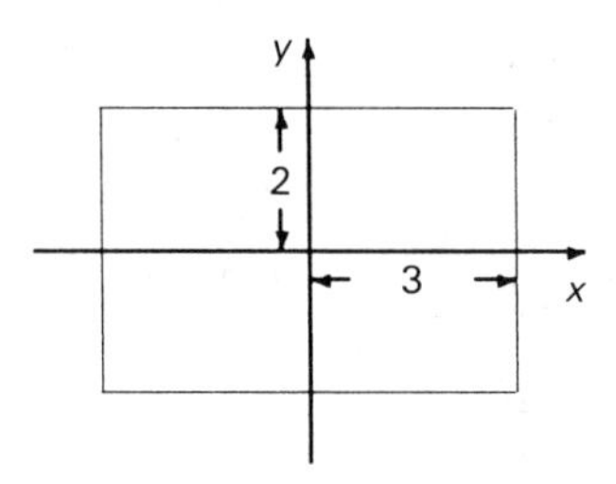

Find the norm, energy, and power for these images. Find also their scalar product and their distance.

The images are digitized using 6×4 identical square fields. First, mid-point sampling in these 24 fields is used; next, averaging over the fields. Find the norm, energy, and power for the resulting digital images.

2.6 Linear approximation of signals

When signals are to be stored, transmitted, or processed in some other way, considerations of economy invariably arise. As mentioned, certain types of signal will be especially suited for a given purpose.

A common practice is the decomposition of the actual signal into a linear combination of other more 'favourable' signals, thus replacing the original signal with its coefficients in the chosen linear combination. The reason is the *linearity* of many signal processing systems occurring in practical applications – a property to be utilized extensively in the remainder of this book. Linearity means that linear combinations are conserved – the coefficients are the same before and after the system action, and the result is described solely in terms of the effect upon the selected signals.

EXAMPLE 2.28

For electromagnetic transmission (alternating current, radio waves, light, etc.), the harmonic signals $f(t) = e^{i\omega t}$ and their linear combinations play a fundamental role, cf. equation (1.11). When processing binary data, for instance when playing a compact disc, the relevant signals are of the rectangular type.

We thus assume that among the various possible signals to be processed, with a given purpose in mind, some are especially suited in some sense; let us denote these signals by $f_1, \ldots, f_K$ and call them *basis signals*.

Thus, if the signal f can be expressed as a linear combination of the K basis signals, that is,

$$f = \sum_{k=1}^{K} a_k f_k \tag{2.25}$$

then the information contained in f is, in principle, reduced to the K coefficients a_k – normally a great advantage, as K typically will be much smaller than the number of samples or pixels.

For good measure, we take care that none of the K basis signals is a linear combination of the others – the inclusion of such a signal would, of course, be superfluous. The set of basis signals thus consists of *linearly independent* signals. In this case, the coefficients a_k in equation (2.25) are unique – the decomposition of f as a linear combination is unambiguous. If $b_1, \ldots, b_K$ were another set of possible coefficients in equation (2.25), that is, if

$$f = \sum_k a_k f_k = \sum_k b_k f_k$$

then

$$\sum_k (a_k - b_k) f_k = 0$$

(where '0' denotes the zero signal, the signal consisting of nothing but zeros). Here, all the differences $(a_k - b_k)$ must be zero. For if one of them, say $(a_l - b_l)$, were not, then

$$f_l = \frac{1}{a_l - b_l} \sum_{k \neq l} (a_k - b_k) f_k$$

in which case the f_l – as a linear combination of the others – would have been excluded from the set.

In general, one should not expect that an arbitrary signal f can be written as a linear combination of the basis signals $f_1, \ldots, f_K$. However:

> There exists precisely one linear combination $f' = \Sigma\, a_k f_k$ which minimizes the distance $|f - f'|$.

For now, our job is to construct a method for determining this linear combination. The problem is a well-known one from geometry – a *projection*. In Figure 2.25, the plane consists of all linear combinations of the vectors f_1 and f_2. The vector f lies outside this plane, but its projection $f' = a_1 f_1 + a_2 f_2$ is precisely the one vector in the plane closest to f – that is, the one for which $|f - f'|$ is minimized.

This vector f' is characterized by $f - f'$ being perpendicular or *orthogonal* to this plane, that is,

$$(f - f') \cdot f_1 = 0 \qquad \text{and} \qquad (f - f') \cdot f_2 = 0$$

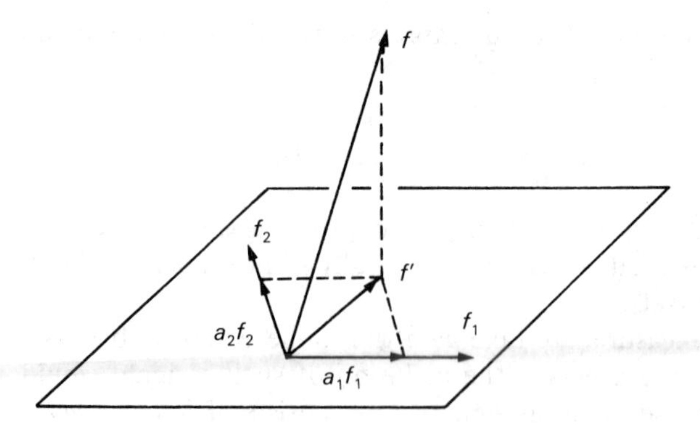

Figure 2.25 Projection of a vector onto a plane.

This projection argument suggests how to solve the problem in the general case where K and N are arbitrary.

If, then, both f' and f'' are linear combinations of the f_ks, but f' satisfies the condition

$$(f - f') \cdot f_k = 0 \qquad \text{for all } k \tag{2.26}$$

then

$$\begin{aligned} |f - f''|^2 &= |(f - f') - (f'' - f')|^2 \\ &= |f - f'|^2 + |f'' - f'|^2 - (f - f') \cdot (f'' - f') - (f'' - f') \cdot (f - f') \end{aligned}$$

In these equalities, $f'' - f'$ is itself a linear combination of the f_ks and, as $f - f'$ is orthogonal to each one of these, the two latter scalar products are zero. The sum thus reduces to

$$|f - f''|^2 = |f - f'|^2 + |f'' - f'|^2$$

which obviously attains its minimum value if we choose $f'' = f'$. So, if the linear combination f' satisfies equation (2.26), it also minimizes the distance to f. If the coefficients in this linear combination are denoted $a_1, \ldots, a_K$, they are found from

$$\sum_{k=1}^{K} a_k f_k \cdot f_l = f \cdot f_l, \qquad l \in [1, K] \tag{2.27}$$

that is, a system of K equations with K unknowns.

This system has exactly one solution $(a_1, \ldots, a_K)$ which may be evaluated by means of one of many suitable methods. However, in one very important case, the solution may be written down without further ado. If the signals f_k are *mutually orthogonal*, that is, if

$$f_k \cdot f_l = 0 \qquad \text{for } k \neq l$$

then the system (2.27) degenerates to $a_k \, | f_k |^2 = f \cdot f_k$, so that the K coefficients become

$$a_k = \frac{f \cdot f_k}{| f_k |^2} \tag{2.28}$$

Orthogonal signals $f_1, \ldots, f_K$ are linearly independent. If, for instance, f_k were a linear combination $f_k = \Sigma_{l \neq k} \, a_l f_l$ of the others, scalar multiplication on both sides of the equation would give

$$f_k \cdot f_k = \sum a_l f_l \cdot f_k = 0$$

The signal f_k would then be the zero signal, which does not belong to the set.

EXAMPLE 2.29

Show that the projection f' of f is always of lesser norm than f and that the two quantities are equal only if $f' = f$.

EXAMPLE 2.30 (Projection in space)

In the usual three-dimensional space, we consider the plane containing the vectors $f_1 = (1, 1, 0)$ and $f_2 = (0, 1, 2)$. We wish to calculate the projection of the vector $f = (1, -1, 5)$ onto this plane.

If this projection is denoted $a_1 f_1 + a_2 f_2$, the coefficients a_1 and a_2 are, according to equation (2.27), to be found from

$$\begin{aligned} (f_1 \cdot f_1) a_1 + (f_2 \cdot f_1) a_2 &= f \cdot f_1 \\ (f_1 \cdot f_2) a_1 + (f_2 \cdot f_2) a_2 &= f \cdot f_2 \end{aligned} \qquad \text{or} \qquad \begin{aligned} 2a_1 - 1a_2 &= 0 \\ 1a_1 + 5a_2 &= 9 \end{aligned}$$

Solution of this system of equations gives $a_1 = -1$, $a_2 = 2$. The result of the projection, thus, is the vector

$$-1(1, 1, 0) + 2(0, 1, 2) = (-1, 1, 4)$$

Find the projection of the vector $(0, 0, 9)$ onto this plane for yourself.

EXAMPLE 2.31 (Linear regression)

One often has to represent a set of measurements $(x_1, y_1), (x_2, y_2), \ldots, (x_N, y_N)$ by a linear relation of the form $y = ax + b$ – a task which, in practice, can never be accomplished exactly. Instead, one often uses the *method of least squares*: finding the values of a and b which minimize the sum of squares $\Sigma \, (y_n - (ax_n + b))^2$ of differences between measured values and 'theoretical values' $ax_n + b$.

One very convenient determination of a and b consists in introducing the digital signals (vectors) $x = (x_1, x_2, \ldots, x_N)$ and $e = (1, 1, \ldots, 1)$, made up of N ones. Then, the job evidently translates into a determination of the linear combination of x and e having minimum distance from $y = (y_1, y_2, \ldots, y_N)$. According to equation (2.27), the solution is found from the system of equations

$$(x \cdot x)a + (e \cdot x)b = x \cdot y$$
$$(x \cdot e)a + (e \cdot e)b = e \cdot y$$

with coefficients

$x \cdot x$	$e \cdot x$	$x \cdot e$	$x \cdot y$	$e \cdot e$	$e \cdot y$
Σx_n^2	Σx_n	Σx_n	$\Sigma x_n y_n$	N	Σy_n

EXAMPLE 2.32

The four analog signals $f_0(t) = 1$, $f_1(t) = t$, $f_2(t) = t^2$ and $f(t) = t^3$, all defined in the interval (0, 1), are sampled equidistantly at the ten points $0, 0.1, 0.2, \ldots, 0.9$. The result is four digital signals $f_0[n], f_1[n], f_2[n], f[n]$ for $n = 0, 1, \ldots, 9$. Find the linear combination of f_0, f_1 and f_2 which best approximates f.

Solution. As, for example, $f_0 \cdot f_0 = 10$, $f_1 \cdot f_2 = 2.025$ and $f \cdot f_2 = 1.208\ 25$, the coefficients in the linear combination sought are found by means of the system of equations

$$\begin{aligned} 10a_0 + 4.5a_1 + 2.85a_2 &= 2.025 \\ 4.5a_0 + 2.85a_1 + 2.025a_2 &= 1.5333 \\ 2.85a_0 + 2.025a_1 + 1.5333a_2 &= 1.208\ 25 \end{aligned}$$

the solution of which is

$$a_0 = 0.0252, \qquad a_1 = -0.461, \qquad a_2 = 1.35$$

What is the result if the fs are considered to be analog signals?

EXAMPLE 2.33 (Digitization and reconstruction)

When an analog signal $f(t)$ is being digitized and reconstructed over the interval $(0, T)$ by division of this interval into N subintervals of length $\Delta t = T/N$, f is simply replaced by a linear combination $\bar{f} = \Sigma\, a_n f_n$ of the N rectangular signals $f_n(t) = r_{\Delta t}(t - n\Delta t)$, $n \in [0, N-1]$. These signals are orthogonal, as

$$f_k \cdot f_l = \int_0^T f_k(t) f_l(t)\, dt = \begin{cases} \Delta t & \text{for } k = l \\ 0 & \text{for } k \neq l \end{cases}$$

Moreover,

$$f \cdot f_k = \int_0^T f(t) r_{\Delta t}(t - k\Delta t)\, dt$$

$$= \int_{k\Delta t}^{(k+1)\Delta t} f(t)\, dt$$

Formula (2.28) accordingly shows that the coefficients a_k in the linear combination which minimizes the distance to f are given by

$$a_k = \frac{1}{\Delta t} \int_{k\Delta t}^{(k+1)\Delta t} f(t)\, dt$$

exactly as in equation (2.6). Thus, averaging over the subintervals (or image fields in the two-dimensional case) is further justified as a digitization principle.

EXAMPLE 2.34

In the interval $(-1, 1)$, two exponential signals $f_-(t) = e^{-t}$ and $f_+(t) = e^t$ are given. Find the linear combination of these which best approximates the signal $f(t) = 1 + t$.

Repeat this exercise in the case where the three analog signals are digitized by, respectively,

(a) sampling in -1, -0.5, 0, 0.5 and 1;
(b) averaging over four identical subdivisions;
(c) mid-point sampling in these intervals, that is, at -0.75, -0.25, 0.25 and 0.75.

2.7 The harmonic orthogonal set

We mentioned in the previous section that the advantage of the linear signal representation lies in the way the representation is conserved, in the sense that signal processing systems do not alter the coefficients appearing in linear combinations. Moreover, it is possible to select signals for which the system effect is simply a multiplication by a constant factor, that is to say, the signal *shape* is unaltered. It is the exponential signals introduced in Section 2.4 which possess this property. Furthermore if the basis signals are chosen to be mutually orthogonal, they will remain orthogonal even after processing. Such orthogonal basis sets will obviously merit special attention.

An important base set of the latter type is the *harmonic orthogonal set*, to be constructed presently. The task will be to specify a set of mutually orthogonal signals of the form $f[n] = z^n$, or, explicitly,

$$f_k = (1, z_k, z_k^2, \ldots, z_k^{N-1}) \tag{2.29}$$

that is, a set of exponential signals on the interval $[0, N-1]$, with possibly complex quotients $z_0, z_1, \ldots$.

It seems natural to ask for the inclusion of the constant signal

$$f_0 = (1, 1, \ldots, 1)$$

in the set. The set is, however, now completely fixed! That is, if $f = (1, z, z^2, \ldots, z^{N-1})$ is included too ($z \neq 1$), then

$$f \cdot f_0 = \sum_{n=0}^{N-1} z^n = \frac{z^N - 1}{z - 1}$$

Thus, the two signals f_0 and f will be orthogonal if this quantity is zero, that is, if z is one of the *complex unit roots* (a solution of the equation $z^N = 1$).

Denoting the simplest one of these w_N,

$$w_N = e^{i \cdot 2\pi/N} \tag{2.30}$$

all the N roots may be expressed as $z_k = w_N^k$, where $k \in [0, N-1]$. Note that $w_N^{N+n} = w_N^n$; the signal $f_k[n] = z_k^n = w_N^{kn}$ will accordingly be *periodic* with period N, if we allow the digital time n to assume all integer values. The signal values in the 'fundamental interval' $[0, N-1]$ are thus repeated both in the past and in the future, and each of the signals f_k consequently possesses only N different values. Also, only N different signals – corresponding to the N different roots $z_k = w_N^k$ – exist. Each one of these is, as shown above, orthogonal to $f_0 = (1, 1, \ldots, 1)$. They are, however, also mutually orthogonal:

$$\begin{aligned} f_k \cdot f_l &= \sum_{n=0}^{N-1} z_k^n (z_l^n)^* \\ &= \sum_{n=0}^{N-1} w_N^{(k-l)n} \\ &= \frac{w_N^{(k-l)N} - 1}{w_N^{(k-l)} - 1} \\ &= 0 \end{aligned} \tag{2.31}$$

for distinct and positive values of k and l. In the case $k = l$, we have

$$|f_k|^2 = \sum w_N^0 = N \tag{2.32}$$

Each of the N signals is, thus, of norm $\sqrt{N}$ and power 1 (cf. p. 54).

EXAMPLE 2.35

By permuting symbols in the above derivation and employing $z_l^k = w_N^{kl} = z_k^l$, show that the δ-signal can be expressed as a linear combination of the basis signals $f_k[n] = z_k^n$:

$$\delta[n-l] = \frac{1}{N} \sum_{k=0}^{N-1} z_k^{-l} z_k^n$$

with $n \in [0, N-1]$ as the digital time and l one of its values.

Summing up:

With the notation $w_N = e^{i \cdot 2\pi/N}$, the N signals $f_k[n] = w_N^{kn}$, where both the index k and the digital time n assume N consecutive integral values, are mutually orthogonal and of norm $\sqrt{N}$.

These signals constitute the harmonic (or trigonometric) orthogonal set of order N. The *normalized* signals $w_N^{kn}/\sqrt{N}$ are referred to as the corresponding harmonic *orthonormal set*. The index k is frequently chosen to lie in the interval $[-M, M]$, if $N = 2M+1$ is odd; for $N = 2M$ one often chooses $k \in [-M, M-1]$.

If we wish to find the best approximation for an arbitrary signal $f[n]$, $n \in [0, N-1]$ with a linear combination of the first K signals in the set, that is, with a signal f' of the form

$$f' = \sum_{k=0}^{K-1} a_k f_k \tag{2.33}$$

the coefficients a_k are, according to equation (2.27),

$$a_k = \frac{1}{N} \sum_{n=0}^{N-1} f[n]\, w_N^{-kn} \tag{2.34}$$

regardless of the value of K. If, in particular, all N signals are employed ($K = N$), the approximation $f \approx f'$ becomes an equality:

$$f = \sum_{k=0}^{N-1} a_k f_k \tag{2.35}$$

The signal f *is* a linear combination of the complete base set (cf. equation (2.10) and Example 2.35) so $f = f'$.

EXAMPLE 2.36

Derive equation (2.35) directly from equation (2.34), performing the summations explicitly.

EXAMPLE 2.37

The signal $f[n]$, $n \in [0, 3]$ is given by

$$f = (0, 1, 0, -1)$$

Approximate this signal with a linear combination f' of the type given by equation (2.33) from the harmonic orthogonal set of order 4. Write the signal f' for all possible cases, that is,

(a) $K = 1$ (b) $K = 2$ (c) $K = 3$ (d) $K = 4$.

Finally, determine the distance $|f - f'|$, thus measuring the quality of the respective approximations.

The analog harmonic orthogonal set consists of the signals

$$f_k(t) = e^{ikt}, \qquad k \in [-\infty, \infty] \tag{2.36}$$

where the index k is an arbitrary integer, positive, negative, or zero. The time interval can be taken to be $(0, 2\pi)$, $(-\pi, \pi)$ or any other, possibly the entire real axis. The set derives from the various digital sets in the limit $N \to \infty$, as the basis signals $f_k[n] = w_N^{kn} = f_k(n \cdot 2\pi/N)$ are N equidistant samples of f_k in the interval $(0, 2\pi)$. The set, which comprises infinitely many analog signals, is orthogonal as expected. This is demonstrated by uniform sampling of the two signals f_k and f_l equidistantly in $(0, 2\pi)$ and forming

$$\sum_{n=0}^{N-1} f_k(n \cdot 2\pi/N) f_l(n \cdot 2\pi/N)^* = \sum_{n=0}^{N-1} w_N^{(k-l)n}$$

As the derivation (2.31) shows, this sum is 0, unless $k - l$ is a multiple of N, and if N increases, the sum will be zero starting from the number $N_0 = k - l + 1$. If we know that $k, l \in [-K, K]$, then the sum is guaranteed to be zero for $N \geqslant 2K + 1$; and the orthogonality in the analog limit now follows from equation (2.22).

EXAMPLE 2.38

Show this orthogonality directly by using equation (2.21), and show that all signals in the set are of norm

$$|f_k| = \sqrt{2\pi} \tag{2.37}$$

If the best approximation of an arbitrary signal f defined on $(-\pi, \pi)$ is to be found with a linear combination from the set, for example based on the $2K + 1$

signals with $k \in [-K, K]$, then the coefficient of f_k will, according to equations (2.28) and (2.21), be given by

$$a_k = \frac{1}{2\pi} \int_{-\pi}^{\pi} f(t) e^{-ikt} \, dt, \qquad k \in [-K, K] \tag{2.38}$$

The most complete reconstruction, the *Fourier series* of f, named after the French mathematician and physicist, J.-B. J. Fourier (1763–1830), is therefore

$$f(t) = \sum_{k=-\infty}^{\infty} a_k e^{ikt} \tag{2.39}$$

in which f is represented by an infinite linear combination of all signals in the analog harmonic orthogonal set.

Since all the signals in the set are periodic with period 2π, any linear combination will have the same property. The Fourier series is thus convenient for the representation of periodic signals, possibly after a change of scale from 2π to the period in question.

EXAMPLE 2.39

Find the Fourier series for the signal $f(t) = t$, $t \in (-1, 1)$ (and for its periodic extensions over $t \in (-\infty, \infty)$), as shown in the figure below.

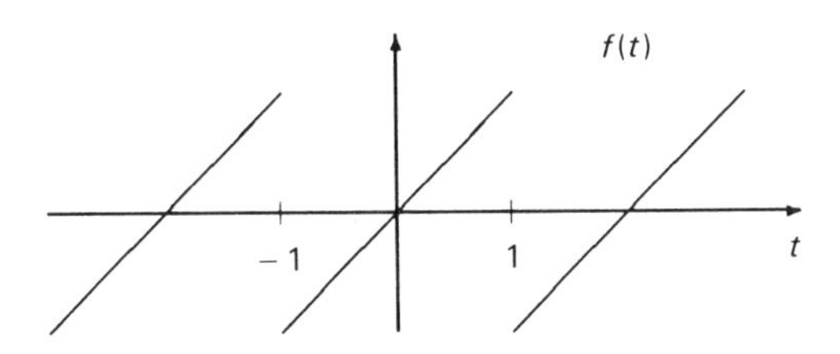

Solution. Had the interval been of length 2π, the solution would have been given by equation (2.38), where, according to equation (2.37),

$$a_0 = 0$$

$$a_k = \frac{1}{2\pi} \int_{-\pi}^{\pi} t e^{-ikt} \, dt = \frac{1}{2\pi} \left[\left(\frac{1}{k^2} + i \frac{t}{k} \right) e^{-ikt} \right]_{-\pi}^{\pi}$$

$$= \frac{1}{2\pi} \left(\left(\frac{1}{k^2} + i \frac{\pi}{k} \right) e^{-ik\pi} - \left(\frac{1}{k^2} - i \frac{\pi}{k} \right) e^{ik\pi} \right)$$

$$= i \frac{(-1)^k}{k} \qquad \text{for } k \neq 0$$

Since $a_k e^{ikt} + a_{-k} e^{i(-k)t} = -2(-1)^k \sin(kt)/k$, the Fourier series becomes

$$t = 2\sum_{k=1}^{\infty} \frac{(-1)^{k+1}}{k} \sin(kt), \qquad t \in (-\pi, \pi)$$

and the simple change of scale from t to πt gives

$$t = \frac{2}{\pi}\sum_{k=1}^{\infty} \frac{(-1)^{k+1}}{k} \sin(k\pi t) \tag{2.40}$$

for the signal $f(t) = t$, $t \in (-1, 1)$.

The equality should be viewed with some caution. The convergence of the infinite series relates to the vanishing difference in signal distance between $f(t)$ and $\Sigma_{k=-K}^{K} a_k e^{ikt}$ as $K \to \infty$. Nevertheless, the series converges pointwise; for instance, equation (2.40) gives, for $t = \frac{1}{2}$,

$$\frac{\pi}{4} = 1 - \frac{1}{3} + \frac{1}{5} - \frac{1}{7} + \cdots$$

whereas the series at the point of discontinuity $t = 1$ consists of zeros, with a total of zero. As will be seen from the figure, at the point $t = 1$, $f(t)$ possesses the two limiting values 1 (from the left) and -1 (from the right), and the convergence towards the *average* $\frac{1}{2}(1 + (-1)) = 0$ in such situations is a quite general property of Fourier series.

Find the Fourier series for the signal $f(t) = |t|$, $t \in (-1, 1)$, for yourself, as well as its value in the four special cases $t = -\frac{1}{2}, 0, \frac{1}{2}$, and 1.

For an arbitrary periodic signal (of period T), the formulae (2.38) and (2.39) should be replaced by

$$f(t) = \sum_{k=-\infty}^{\infty} a_k e^{ik\omega t}, \qquad \text{with} \qquad a_k = \frac{1}{T}\int_0^T f(t) e^{-ik\omega t}\, dt \tag{2.41}$$

where $\omega = 2\pi/T$ is the angular frequency of the signal.

If a signal $f(t)$ – say, once again defined on the interval $(0, 2\pi)$ – is to be represented in practice by (part of a) Fourier series, it is never specified in a closed, functional form, for which the integration given in equation (2.38) can be effected. More likely, a number of samples $f(t_n)$ will be given, and from the considerations on pp. 63–4 it follows that, from N samples, one is able to calculate exactly N Fourier coefficients a_k, $k \in [0, N-1]$, so that

$$f(t_n) = \sum_{k=0}^{N-1} a_k e^{ikt_n}$$

if the sampling takes place at $t_n = n \cdot 2\pi/N$. This means that the relation

$$f(t) = \sum_{k=0}^{N-1} a_k e^{ikt} \tag{2.42}$$

with coefficients a_k given by equation (2.34), *interpolates* f: it reproduces f at the sampling points t_n and suggests values for f at the remaining points.

EXAMPLE 2.40 (The sampling theorem)
A real signal $f(t)$, $t \in (-\pi, \pi)$, is specified by virtue of its samples $f[n] = f(n \cdot \pi/M)$, $n \in [-M, M]$. We shall interpolate it by means of the expression

$$f(t) = \sum_{k=-M}^{M} a_k e^{ikt} = a_0 + \sum_{k=1}^{M} (a_k e^{ikt} + a_{-k} e^{-ikt})$$

The latter sum consists of the terms

$$\begin{aligned} a_k e^{ikt} + a_{-k} e^{-ikt} &= a_k(\cos kt + i \sin kt) + a_{-k}(\cos kt - i \sin kt) \\ &= (a_k + a_{-k})\cos kt + i(a_k - a_{-k})\sin kt \end{aligned}$$

As will be seen from equation (2.33), $a_{-k} = a_k^*$, so that

$$f(t) = a_0 + \sum_{k=1}^{M} (b_k \cos kt + c_k \sin kt)$$

where b_k and $-c_k$ denote the real and imaginary parts, respectively, of $2a_k$:

$$2a_k = b_k - ic_k, \qquad k \in [1, M]$$

If f is of period T instead of 2π, its representation will be

$$f(t) = a_0 + \sum_{k=1}^{M} \left[b_k \cos\left(\frac{2\pi}{T} kt\right) + c_k \sin\left(\frac{2\pi}{T} kt\right) \right] \tag{2.43}$$

with

$$\begin{aligned} b_k &= \frac{1}{T} \int_0^T f(t) \cos\left(\frac{2\pi}{T} kt\right) dt \\ c_k &= \frac{1}{T} \int_0^T f(t) \sin\left(\frac{2\pi}{T} kt\right) dt \end{aligned} \tag{2.44}$$

Since f is sampled with frequency $2M/T$, equation (2.43) may be said to express the fact that the signal can be reconstructed or 'synthesized' from real harmonic signals having frequencies from 0 all the way up to the *half sampling frequency* (M/T, corresponding to the angular frequency $2\pi M/T$), the so-called *Nyquist frequency* for this case. If, on the other hand, a signal consisting of harmonic components has been sampled with less than double the highest frequency occurring, high-frequency components will erroneously be identified with low-frequency ones. This phenomenon is known as *aliasing*.

EXAMPLE 2.41
A signal $f(t)$ in $(-20, 25)$ has been sampled as listed in the table below (see also Figure 2.12).

t	−20	−15	−10	−5	0	5	10	15	20	25
f	10	8.5	6	0	−2.5	−4	−2	0	3	10

This signal is to be reconstructed using an expression of the type given by equation (2.43) with $M=4$. By this means, find $f(2.5)$ and $f(7.5)$.

If two signals f and g have been specified by their individual Fourier coefficients a_k and b_k, $k \in [-\infty, \infty]$, one has

$$f \cdot g = \left(\sum_k a_k f_k\right) \cdot \left(\sum_l b_l f_l\right)$$

$$= \sum_k \sum_l a_k b_l^* f_k \cdot f_l$$

$$= 2\pi \sum_{k=-\infty}^{\infty} a_k b_k^* \tag{2.45}$$

In particular,

$$|f|^2 = 2\pi \sum_{k=-\infty}^{\infty} |a_k|^2 \tag{2.46}$$

EXAMPLE 2.42
Using Example 2.39 and equation (2.46), show that

$$\frac{1}{1^2} + \frac{1}{2^2} + \frac{1}{3^2} + \cdots = \frac{\pi^2}{6}$$

In the image case, the harmonic orthogonal sets are made up of products of one-dimensional harmonic orthogonal signals:

$$f_{kl}[m, n] = w_M^{km} w_N^{ln}$$

$$= e^{i \cdot 2\pi(km/M + ln/N)} \tag{2.47}$$

where $m \in [0, M-1]$ and $n \in [0, N-1]$ are digital position coordinates. With the definition given in equation (2.20) of the scalar product we conclude that all images in the set are of norm

$$|f_{kl}| = \sqrt{MN} \tag{2.48}$$

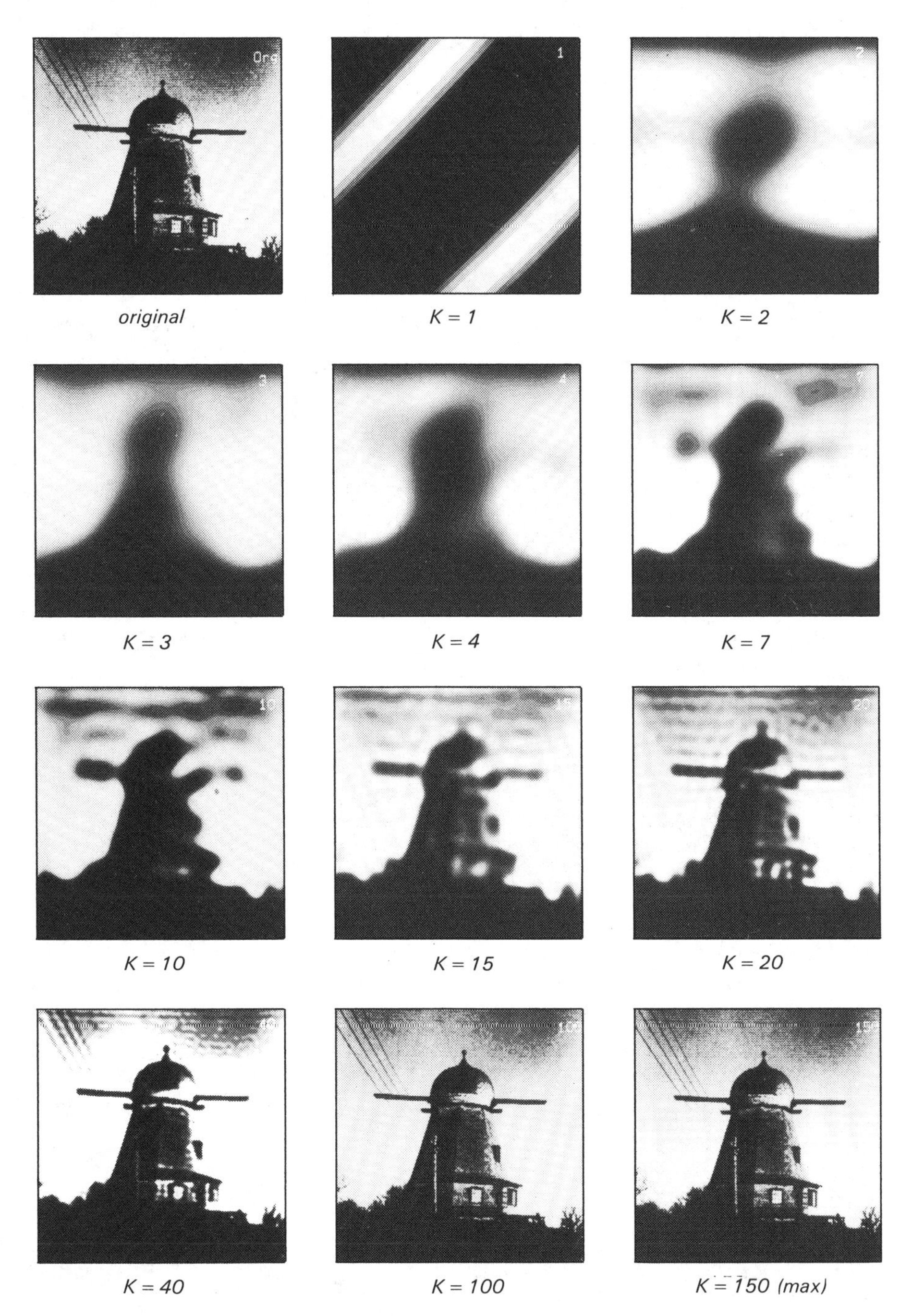

Figure 2.26 Harmonic image approximations.

If, now, an image b is to be approximated optimally by a linear combination $\Sigma_k \Sigma_l a_{kl} f_{kl}[m,n]$, the coefficients are

$$a_{kl} = \frac{1}{MN} \sum_m \sum_n b[m,n] \, w_M^{-km} w_N^{-ln} \tag{2.49}$$

In the analog limit, the harmonic orthogonal images are

$$f_{kl}(s,t) = \mathrm{e}^{iks} \mathrm{e}^{ilt} \tag{2.50}$$

with $s, t \in (0, 2\pi)$ and $k, l \in [-\infty, \infty]$. Each of these images is of norm

$$|f_{kl}| = 2\pi \tag{2.51}$$

Here, too, equation (2.49) should be used to calculate the coefficients both when approximating and interpolating images by means of (products of) Fourier series.

EXAMPLE 2.43

In Figure 2.26 (see previous page), the image in the upper left-hand corner has been approximated by linear combinations of signals from the set f_{kl}, where k and l both run from $-K$ to K. The effect of increasing K, that is, of including more and more high-frequency (spatial) components, is obvious, as is the possibility of representing the large original information content (64 levels and $256 \times 256 = 65\,536$ pixels) by a much smaller number of Fourier coefficients.

3

Transformations and Systems

Transformations, including the important subclass called 'systems', are manipulations which alter signals or represent them in a new form. For instance, we saw in Section 2.6 how to write a signal as a linear combination of basis signals; in this case, the set of coefficients constitutes a 'transform' of the original signal.

An essential motivation for the introduction of transforms is, as usual, economy. If a digital image is to be transmitted or processed, the organization of its information will be of the utmost importance. This is where the transform comes in.

The philosophy will become clear from the simple example of an image containing only a very modest amount of information (Figure 3.1); however, this information is represented in a very inconvenient form. Consisting of only two levels $g_0 = 0$ and $g_1 = 1$, the image is obviously non-separable. If, however, the objects are rotated so as to make their edges parallel to those of the image frame (or rather, if the frame is oriented in the same way as the objects), separability ensues, with an associated reduction of the information content.

Intuitively, it appears highly unsatisfactory that one should be able to remove the amount of information in an image just by rotating it – or by turning one's head in the right direction! To remedy this and similar situations, this chapter is devoted to changes in the representation of image information – *transformations*.

3.1 Transformations

A transformation is a means of altering signals into other signals. Transformations conserve linear combinations; they are said to be *linear*. The various possible signals acted upon, or to be acted upon, are called *inputs* (or sometimes, *stimuli*). The transformed signals are called *outputs* (or *responses*).

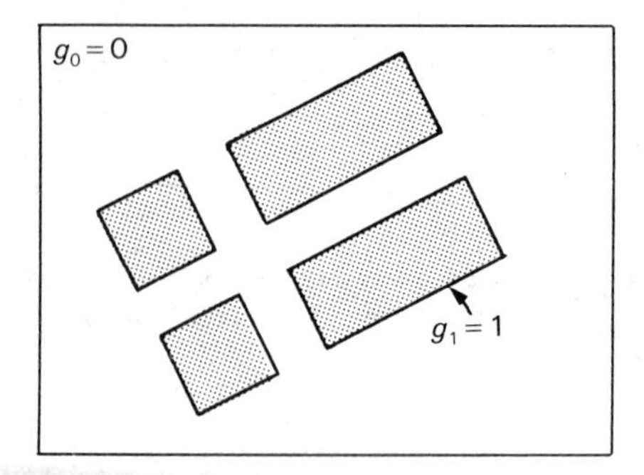

Figure 3.1 Image with rectangular objects.

If a transformation T alters the signal f into the signal g, this is expressed in the simple notation

$$g = Tf \tag{3.1}$$

The action of T is thus denoted by the letter preceding the signal; the notation is shorthand for the more explicit statement $g[n] = Tf[n]$ *for all* n. For two transformations T and U to be identical, they must have the same effect on all possible input signals; that is, $T = U$ means

$$Tf[n] = Uf[n] \text{ for all } f \text{ and all } n$$

In principle, each value $Tf[n]$ of the output will depend on the totality of input signal values $f[n]$. The complete specification of the signal action could, in general, become quite complex, unless *linearity* were there to assist us. The linearity of T is expressed as follows:

$$T\left(\sum a_k f_k\right) = \sum a_k T f_k \tag{3.2}$$

for all complex numbers a_k and all input signals f_k. On the right-hand side of this equation, T acts upon a signal which is a linear combination of the signals f_k, the coefficients being a_k. The result will be the *same* linear combination, now of the signals Tf_k. This property may be expressed by saying that transformations and *linear operations* (addition of two or more signals, multiplication of a signal by some coefficient) are *commutable* – the order of performance of these manipulations is irrelevant.

EXAMPLE 3.1
Signals $f[n]$ are altered into new signals $g[n]$ as follows:

$$g[n] = f[n-3] + 2f[n] - f[n+1]$$

This rule is linear, that is, a transformation. If, however, $f[n]$ is altered into $f[n]+1$, $f[n]^2$, or $|f[n]|$, the transformation property is not satisfied.

Which of the signal changes below (where f is altered into g) are linear?

(a) $g[n] = f[n] + n$ (b) $g[n] = nf[n]$ (c) $g[n] = n^2 f[n]$

One of the simplest examples of a transformation is

$$Tf[n] = af[n] \tag{3.3}$$

a being a complex number. The transformation T so defined is called an *amplification* with *gain* a (Figure 3.2).

Another important class of transformations are the *translations* T_k defined by

$$T_k f[n] = f[n-k] \tag{3.4}$$

where k is an integer (Figure 3.3). The action consists of a shift of the input k units forward in time (backward if k is negative). The translation T_1 is sometimes called the *unit delay element*.

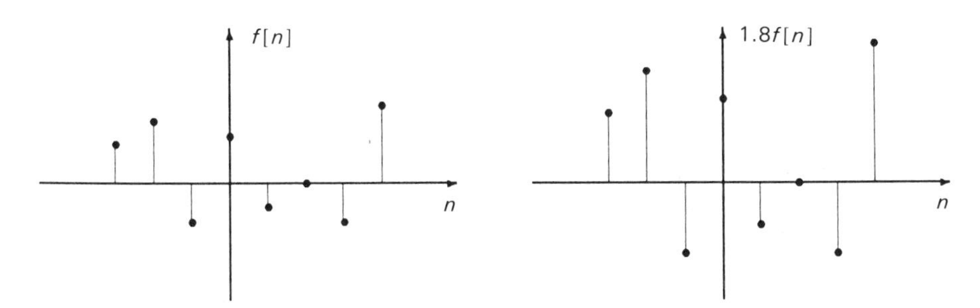

Figure 3.2 Amplification of a signal.

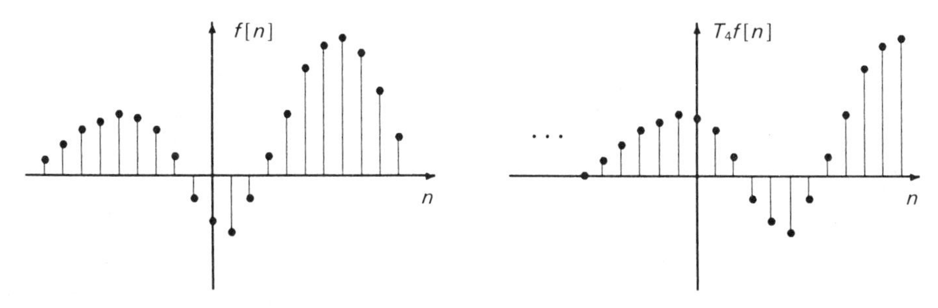

Figure 3.3 Translation of a signal.

EXAMPLE 3.2

The input signal f is defined by $f[n] = n^2 + 1$. Find the following output values:

(a) $T_1 f[0]$ (b) $T_{-3} f[2]$ (c) $T_2 f[-2]$.

Translations of images are defined analogously (Figure 3.4). Image translations must be specified by means of two indices corresponding to the shift in horizontal and vertical position, respectively:

$$T_{kl} b[m, n] = b[m - k, n - l] \tag{3.5}$$

Once again, it is stressed that k and l are parameters denoting specific image coordinates, while m and n are part of the symbolic signal notation for the image.

3.1.1 Composite transformations

If a signal undergoes two transformations in succession – that is, the output of the first transformation serves as the input to the second – the two transformations are said to be *composed* into a new transformation.

The resulting transformation, the *composite* of the two transformations T and U, is simply denoted UT (note the inverted form, in accordance with expression (3.1)), and is defined as follows:

$$(UT)f = U(Tf) \tag{3.6}$$

This tells us how the effect of the composite UT upon the input f (the left-hand side of the equation) is obtained. First, let T act upon f, resulting in Tf; next, let U act upon Tf (the right-hand side).

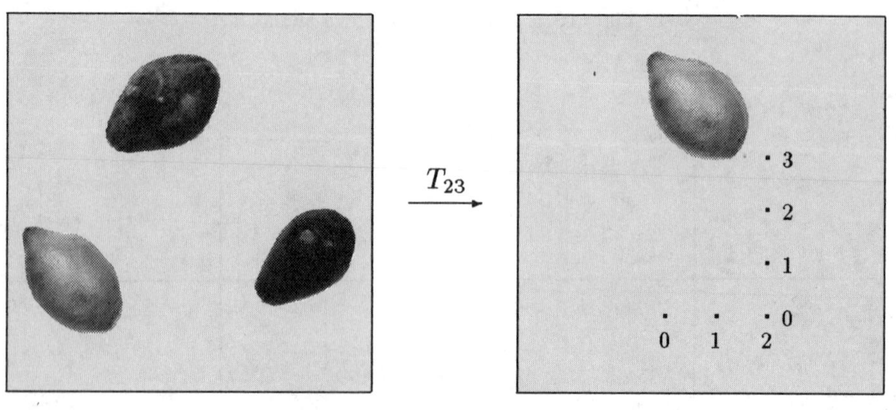

Figure 3.4 Translation of an image.

It is *very* important to realize that, in general, UT is *not* the same transformation as TU, as illustrated by the following two examples.

EXAMPLE 3.3

If the transformation T is given by $Tf[n] = f[n-1]$, and U by $Uf[n] = nf[n]$, one has

$$UTf[n] = nf[n-1]$$
$$TUf[n] = (n-1)f[n-1]$$

which are obviously different results.

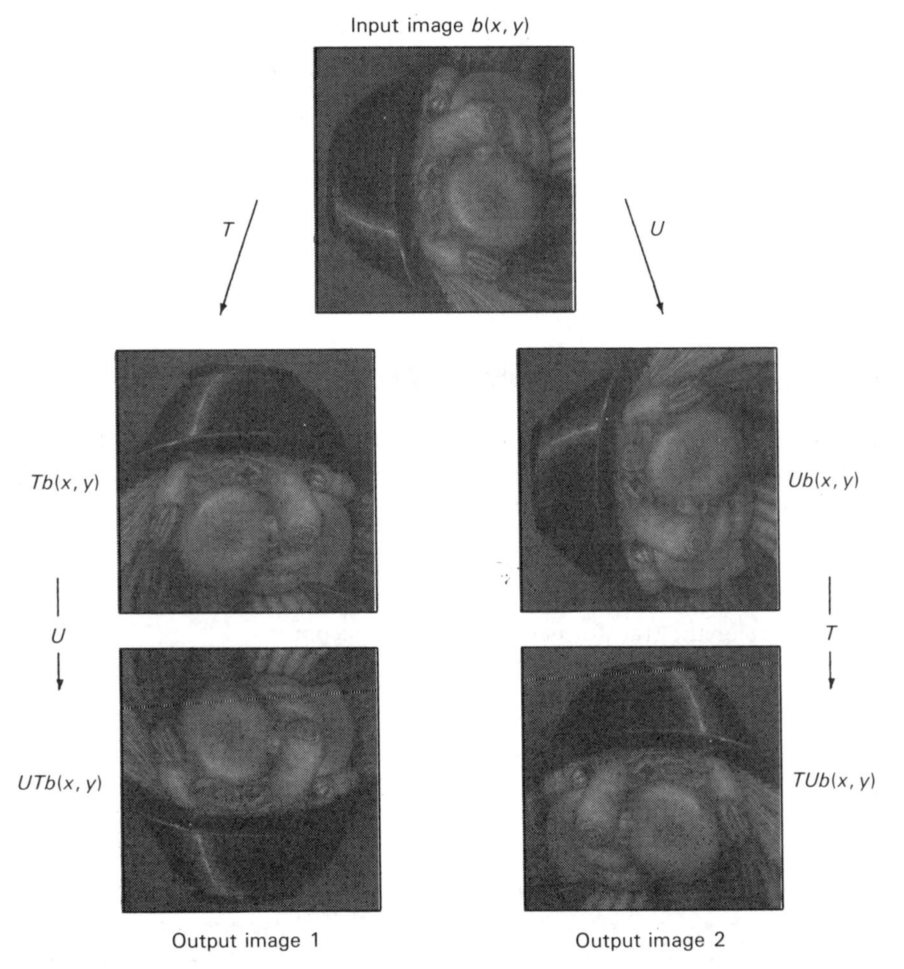

Figure 3.5 Effect of interchanging two transformations.

Find the values $UTf[n]$ and $TUf[n]$, for $f[n] = 2n^2 + 1$ and the three digital time values $n = -1$, 0 and 1.

EXAMPLE 3.4

In Figure 3.5 (see p. 75), T is the transformation rotating a square image 90° (clockwise) about its centre, while the action of U is to form a mirror image about the horizontal axis of symmetry. The effects of the two individual transformations as well as of the composite transformations are shown. The end results are not identical.

On the other hand, many important cases exist (cf. Section 3.4) where the transformations T and U are commutable, that is, where $TU = UT$: $TUf[n] = UTf[n]$ for all inputs f and all points in time n. The translations provide an immediate example. Clearly, T_k may be achieved by composing T_1 with itself k times; more generally,

$$T_k T_l = T_{k+l} \tag{3.7}$$

Thus translations are mutually interchangeable.

3.2 Characterization of transformations

The conservation of linear combinations provides a convenient means of specifying the effect of a transformation completely. An arbitrary signal f can, as noted, be expressed as a linear combination:

$$f[n] = \sum_k f[k]\,\delta[n-k] \tag{3.8}$$

and the effect of the transformation T upon f becomes

$$Tf[n] = \sum_k f[k]\,T(\delta[n-k])$$

This relation fixes the effect of T, if only the effect upon the various time-translated δ-signals is known. If these resulting signals are denoted $h_k[n] = T(\delta[n-k])$, we obtain

$$Tf[n] = \sum_k f[k]\,h_k[n] \tag{3.9}$$

The transformation T is thus completely characterized by the signals $h_k[n]$, the so-called *impulse responses* of T – the outputs corresponding to δ-signal inputs.

The action of a given transformation T having impulse responses $h_k[n]$ therefore consists in *using the values of the input signal as coefficients in this basis set*; the output is, then, the linear combination thus formed.

At the same time, this characterization of T leads to a method for *inverting* a transformation T. Given a transformation T, which *inverse* transformation T^{-1} has the effect of cancelling T? That is, we wish to find T^{-1} such that

$$T^{-1}(Tf) = f \qquad \text{for all inputs } f$$

That is, given a signal g $(= Tf)$, we only have to write it as a linear combination of the signals $h_k[n] = T(\delta[n-k])$; to this end, one may employ equation (2.27) or (2.28). The resulting coefficients then simply equal $T^{-1}g[k]$ $(= f[k])$.

In conclusion, we may state that, in a transformation, signals are replaced by their coefficients in various linear combinations of different basis sets. Here, the signals $f_k[n] = \delta[n-k]$ enjoy a unique status, as the signal coefficients in this basis set equal the original signal values (cf. equation (3.8)).

EXAMPLE 3.5
If T^{-1} has impulse responses $H_l[n]$, then (cf. Example 2.35)

$$\delta[n-l] = \sum_k H_l[k]\, h_k[n]$$

EXAMPLE 3.6
An amplifier of gain a is characterized by its impulse responses $h_k[n] = a\delta[n-k]$. The translation T_l has $h_k[n] = \delta[n-k-l]$.

EXAMPLE 3.7 (Rotation in the plane)
If the vectors in the plane shown in Figure 3.6 are rotated anticlockwise by an angle θ, this transformation is represented by

$$h_1 = (\cos\theta, \sin\theta)$$
$$h_2 = (-\sin\theta, \cos\theta)$$

EXAMPLE 3.8
A transformation T in space changes the basis vectors (1, 0, 0), (0, 1, 0), and (0, 0, 1) into (1, 1, 0), (0, 1, 1), and (1, −1, 5), respectively. What happens to the basis vectors when the inverse transformation T^{-1} is applied?

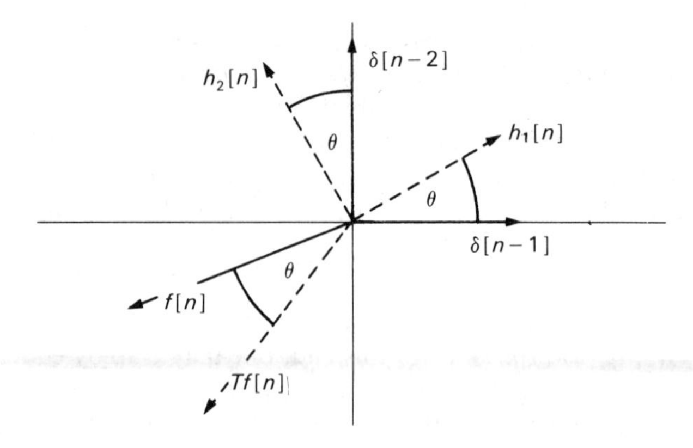

Figure 3.6 Vectors in the plane (signal notation).

3.3 The Fourier transform

The Fourier transform is, possibly, the most powerful tool in signal analysis. Its content is, however, extremely simple. Signals are represented as linear combinations of basis signals from the harmonic orthonormal set. A signal is said to be *resolved* into its harmonic components.

First, we examine the digital Fourier transform. Then, the analog version, as usual, results from increasingly fine sampling of the analog signals in question.

3.3.1 The digital Fourier transform

As mentioned, the digital Fourier transform employs the harmonic orthonormal set

$$f_m[n] = \frac{1}{\sqrt{N}} w_N^{mn}, \qquad m, n \in [0, N-1] \tag{3.10}$$

If a digital signal $f[n]$ is to be represented as a linear combination $f = \Sigma_m F[m] f_m$, that is,

$$f[n] = \frac{1}{\sqrt{N}} \sum_{m=0}^{N-1} F[m]\, w_N^{mn} \tag{3.11}$$

the coefficients $F[m]$ are found from

$$F[m] = \frac{1}{\sqrt{N}} \sum_{n=0}^{N-1} f[n]\, w_N^{-mn} \tag{3.12}$$

with reference to equation (2.34). The signal consisting of these N values $F[m]$ is called the (direct) *digital Fourier transform* (DFT), of the signal $f[n]$.

Correspondingly, equation (3.11) is the *inverse* DFT – the signal $f[n]$ is, then, the inverse DFT of the signal $F[m]$. In this context, the integer N is referred to as the *order* of the Fourier transform.

The inverse Fourier transform is, thus, a Fourier transform itself (corresponding, however, to a reversed order of the orthonormal signals) – one among several useful properties of this important class of transformations.

EXAMPLE 3.9

We compute the DFT for the signal

$$f[n] = (1, -\mathrm{i}, 0, 2\mathrm{i})$$

The simple complex unit root of order 4 is i. Hence,

$$\begin{aligned}
F[0] &= (1\cdot \mathrm{i}^{-0\cdot 0} + (-\mathrm{i})\cdot \mathrm{i}^{-0\cdot 1} + 0\cdot \mathrm{i}^{-0\cdot 2} + 2\mathrm{i}\cdot \mathrm{i}^{-0\cdot 3})/\sqrt{4} \\
&= (1 - \mathrm{i} + 0 + 2\mathrm{i})/2 \\
&= \tfrac{1}{2}(1 + \mathrm{i}) \\
F[1] &= (1\cdot \mathrm{i}^{-1\cdot 0} + (-\mathrm{i})\cdot \mathrm{i}^{-1\cdot 1} + 0\cdot \mathrm{i}^{-1\cdot 2} + 2\mathrm{i}\cdot \mathrm{i}^{-1\cdot 3})/\sqrt{4} \\
&= (1 - 1 + 0 - 2)/2 \\
&= -1 \\
F[2] &= (1\cdot \mathrm{i}^{-2\cdot 0} + (-\mathrm{i})\cdot \mathrm{i}^{-2\cdot 1} + 0\cdot \mathrm{i}^{-2\cdot 2} + 2\mathrm{i}\cdot \mathrm{i}^{-2\cdot 3})/\sqrt{4} \\
&= (1 + \mathrm{i} + 0 - 2\mathrm{i})/2 \\
&= \tfrac{1}{2}(1 - \mathrm{i}) \\
F[3] &= (1\cdot \mathrm{i}^{-3\cdot 0} + (-\mathrm{i})\cdot \mathrm{i}^{-3\cdot 1} + 0\cdot \mathrm{i}^{-3\cdot 2} + 2\mathrm{i}\cdot \mathrm{i}^{-3\cdot 3})/\sqrt{4} \\
&= (1 + 1 + 0 + 2)/2 \\
&= 2
\end{aligned}$$

Find the DFT of the signal $(1 + \mathrm{i}, -2, 1 - \mathrm{i}, 4)$ for yourself.

For clarity, we repeat the important formulae (3.11) and (3.12) constituting the digital Fourier transform in Table 3.1. The DFT expresses the fact that if the signal f is represented by F in the basis set $w_N^{mn}/\sqrt{N}$, then F is represented by f in the basis set $w_{\bar{N}}^{mn}/\sqrt{N}$.

Table 3.1 The digital Fourier transform of order N.

Direct DFT	Inverse DFT
$F[m] = \dfrac{1}{\sqrt{N}} \sum_{n=0}^{N-1} f[n]\, w_{\bar{N}}^{mn}$	$f[n] = \dfrac{1}{\sqrt{N}} \sum_{m=0}^{N-1} F[m]\, w_N^{mn}$

EXAMPLE 3.10

Find the inverse DFT of the signal $(1+\mathrm{i}, -2, 1-\mathrm{i}, 4)$, and check the result using Example 3.9.

As will be seen in the next section, signal processing systems possess the property of conserving (or rather amplifying) harmonic signals, so that each component in the processed signal is merely multiplied by a constant. This means that the processed signal may immediately be written in its Fourier-transformed version, after which it may be presented in its 'direct' version after an inverse Fourier transformation.

Next, a few results concerning the connection between Fourier transformation and summation:

$$\sum_{n=0}^{N-1} f[n] = \sqrt{N}F[0] \qquad \text{and} \qquad \sum_{m=0}^{N-1} F[m] = \sqrt{N}f[0] \tag{3.13}$$

$$f \cdot g = F \cdot G \tag{3.14}$$

Here, F and G are DFTs of the signals f and g, respectively; $F \cdot G = \Sigma_m F[m]G[m]^*$ is their scalar product. From equation (3.14) it follows, in particular, that

$$|F| = |f| \tag{3.15}$$

The formula (3.13) is immediately obtained from Table 3.1, and equation (3.14) is shown to hold as follows:

$$\begin{aligned} f \cdot g &= \left(\sum_k F[k] f_k\right) \cdot \left(\sum_l G[l] f_l\right) \\ &= \sum_k \sum_l F[k] G[l]^* f_k \cdot f_l = \sum_k F[k] G[k]^* \\ &= F \cdot G \end{aligned}$$

Finally, as regards digital Fourier transforms of images $b[m,n]$, $m \in [0, M-1]$, $n \in [0, N-1]$, the natural generalization is

$$B[r,s] = \frac{1}{\sqrt{MN}} \sum_{n=0}^{N-1} \sum_{m=0}^{M-1} b[m,n]\, w_M^{-rm} w_N^{-sn} \tag{3.16}$$

This expression is the two-dimensional DFT for b. Note that the inner sum (over m) is the usual (one-dimensional) DFT, since n, here, is fixed – that is, a parameter. For each n, the result is a DFT which is subsequently transformed along n.

In practice, the computation of an image DFT is carried out in two steps, the first being a transform of each row (or column) in the image, the second a transformation of each column (or row) in the 'intermediate' image.

It is easy to demonstrate that the inverse two-dimensional DFT is given by the following expression:

$$b[m,n] = \frac{1}{\sqrt{MN}} \sum_{n=0}^{N-1} \sum_{m=0}^{M-1} B[r,s]\, w_M^{rm} w_N^{sn} \tag{3.17}$$

EXAMPLE 3.11
If $b[m,n]$ is an image, we employ the notation

$$b[,n] = \sum_{m=0}^{M-1} b[m,n]$$

for the one-dimensional signal resulting from a summation over the rows in b. Show, using equation (3.16), that $B[,s]/\sqrt{M}$ is the Fourier transform of $b[,n]$.

If b is separable, that is, if $b[m,n] = c[m]\,d[n]$, the two-dimensional transform is simply the product of two one-dimensional transforms, that is, $B[r,s] = C[r]D[s]$. In general, this decomposition results in considerable computational advantages.

EXAMPLE 3.12
Find the (two-dimensional) DFT for the following images:

(a)

1	0	1
0	1	0
1	0	1

(b)

1	−1	1
−1	1	−1
1	−1	1

3.3.2 The analog Fourier transform

In Section 2.3, we mentioned the two types of analog signal whose information content was localized within finite intervals of time – periodic signals and transient signals. Moreover, we have seen (Section 2.7) how periodic signals could be expressed as linear combinations of 'pure' harmonic signals (signals of the form $f(t) = e^{i\omega t}$). Here, the relevant values of ω were all multiples of the fundamental frequency of the original signal, ω_0, that is, $\omega = n\omega_0$, $n \in [-\infty, \infty]$. In the present section, we focus upon the transient signals, which similarly may be resolved into harmonic signals of all possible frequencies ω, that is, $f(t) = e^{i\omega t}$, with $\omega \in (-\infty, \infty)$. We therefore consider a signal like that in Figure 3.7. The signal f is now sampled in the $N = 2M$ points $n \in [-M, M-1]$, and the resulting digital signal $f[n]$ is

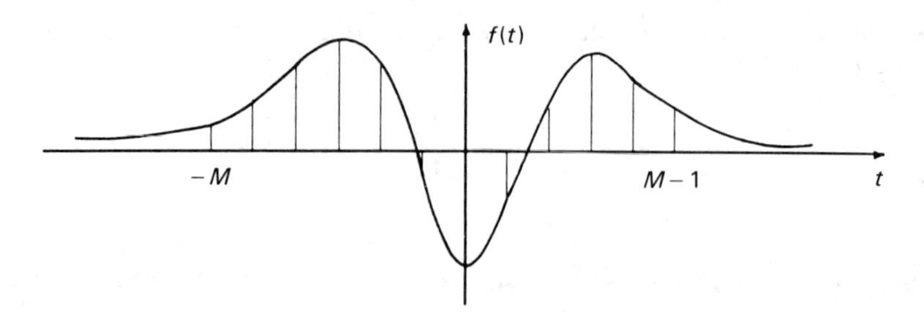

Figure 3.7 A transient signal.

Fourier-transformed digitally as shown in Table 3.1. In order to take $N\to\infty$, these two relations are shown like this (cf. equation (2.22)):

$$\frac{F[m]}{\sqrt{N}}=\frac{1}{N}\sum_n f[n]\,\mathrm{e}^{-\mathrm{i}2\pi mn/N} \qquad \text{and} \qquad \frac{f[n]}{\sqrt{N}}=\frac{1}{N}\sum_m F[m]\,\mathrm{e}^{\mathrm{i}2\pi mn/N}$$

Next, we regard all the signals appearing here as being samples of analog signals, as summarized in Table 3.2. Then, the formula (2.22) shows that in the analog limit

$$\frac{F[m]}{\sqrt{N}}=\frac{1}{T}\int_{-T/2}^{T/2} f(t)\mathrm{e}^{-\mathrm{i}2\pi mt/T}\,\mathrm{d}t \tag{3.18}$$

$$\frac{f[n]}{\sqrt{N}}=\frac{1}{\Omega}\int_{-\Omega/2}^{\Omega/2} F(\omega)\mathrm{e}^{\mathrm{i}2\pi n\omega/\Omega}\,\mathrm{d}\omega \tag{3.19}$$

Examining equation (3.18) first, we can determine $F(\omega)$ by choosing m as near to $N\omega/\Omega$ as possible:

$$\frac{F(\omega)}{\sqrt{N}}=\frac{1}{T}\int_{-T/2}^{T/2} f(t)\mathrm{e}^{-\mathrm{i}2\pi(N\omega/\Omega)t/T}\,\mathrm{d}t \tag{3.20}$$

In order to ensure that all the information in $f(t)$ is included, we now let T increase towards ∞. However, to obtain an increasing fine sampling, it is necessary to make T grow more slowly than N – for instance, as $\sqrt{N}$. Further, one may choose Ω

Table 3.2 Analog signals and samples for Fourier transformation.

Interval	$(-T/2, T/2)$		$(-\Omega/2, \Omega/2)$	
Sampling points	$t=nT/N$		$\omega=m\Omega/N$	
Digital signal	$f[n]$	$\mathrm{e}^{-\mathrm{i}2\pi mn/N}$	$F[m]$	$\mathrm{e}^{\mathrm{i}2\pi mn/N}$
Analog signal	$f(t)$	$\mathrm{e}^{-\mathrm{i}2\pi mt/T}$	$F(\omega)$	$\mathrm{e}^{\mathrm{i}2\pi n\omega/\Omega}$

freely, and with reference to equation (3.20), $\Omega T = 2\pi N$ seems convenient. For the sake of symmetry, we take

$$\Omega = T = \sqrt{2\pi N}$$

In the analog limit $N \to \infty$, we thus obtain from equation (3.20)

$$F(\omega) = \frac{1}{\sqrt{2\pi}} \int_{-\infty}^{\infty} f(t) e^{-i\omega t} \, dt \tag{3.21}$$

This expression is the *analog (direct) Fourier transform*. Turning next to equation (3.15), if we similarly put $n = Nt/T$, we obtain

$$\frac{f(t)}{\sqrt{N}} = \frac{1}{\Omega} \int_{-\Omega/2}^{\Omega/2} F(\omega) e^{i2\pi(Nt/T)\omega/\Omega} \, d\omega$$

that is,

$$f(t) = \frac{1}{\sqrt{2\pi}} \int_{-\infty}^{\infty} F(\omega) e^{i\omega t} \, d\omega \tag{3.22}$$

which is the *inverse Fourier transform*. Equations (3.21) and (3.22) together constitute the analog Fourier transformation, allowing a decomposition of a transient signal into harmonic components $e^{i\omega t}$ with (complex) amplitudes $F(\omega)$. A signal $f(t)$ is thus completely characterized by its *amplitude spectrum* $|F(\omega)|$ and its *phase spectrum* $\phi(\omega)$.

The two complementary formulae are repeated in Table 3.3.

EXAMPLE 3.13

The Fourier transform of the rectangular signal $f(t) = r_T(t - T/2)$ (Figure 3.8a) is given by

$$\begin{aligned} F(\omega) &= \frac{1}{\sqrt{2\pi}} \int_{-\infty}^{\infty} f(t) e^{-i\omega t} \, dt \\ &= \frac{1}{\sqrt{2\pi}} \int_{-T/2}^{T/2} e^{-i\omega t} \, dt \\ &= \frac{1}{\sqrt{2\pi}} \frac{1}{-i\omega} [e^{-i\omega T/2} - e^{i\omega T/2}] \\ &= \sqrt{\frac{2}{\pi}} \frac{\sin(\omega T/2)}{\omega} \end{aligned}$$

(Figure 3.8b). The latter function is of fundamental importance in signal analysis; with the notation $\text{sinc}(t) = \sin t/t$ (*sinus cardinalis*), we thus have

$$F(\omega) = \frac{1}{\sqrt{2\pi}} T \,\text{sinc}\left(\frac{T}{2} \omega\right) \tag{3.23}$$

Table 3.3 Analog Fourier transformation.

Direct Fourier transform	Inverse Fourier transform
$F(\omega)=\frac{1}{\sqrt{2\pi}}\int_{-\infty}^{\infty} f(t)e^{-i\omega t}\,dt$	$f(t)=\frac{1}{\sqrt{2\pi}}\int_{-\infty}^{\infty} F(\omega)e^{i\omega t}\,d\omega$

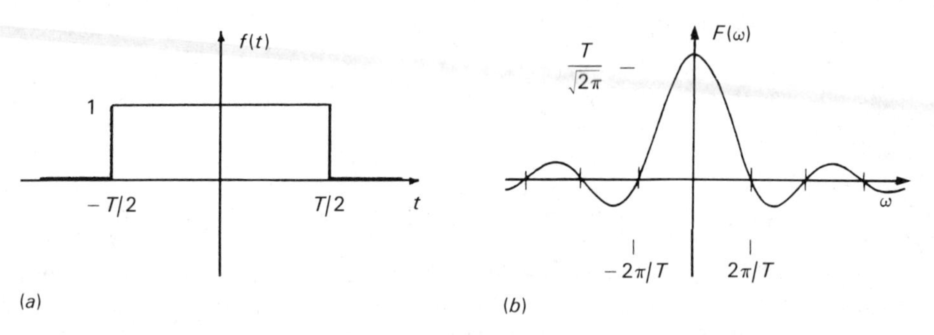

Figure 3.8 The rectangular signal and its Fourier transform.

In most situations, $f(t)$ is a real signal, whereas $F(\omega)$ is a complex one, since $e^{-i\omega t} = \cos \omega t - i \sin \omega t$ is complex. The following decomposition

$$F(\omega)=\frac{1}{\sqrt{2\pi}}\int f(t)\cos(\omega t)\,dt - i\,\frac{1}{\sqrt{2\pi}}\int f(t)\sin(\omega t)\,dt \tag{3.24}$$

(in which the two components are called the *cosine* and *sine transforms* of f) will in this case consist of the real and imaginary parts of the (complex) Fourier transform F. Moreover, if f is of either *even* or *odd parity*, that is, if either $f(t)=f(-t)$ (Figure 3.9a) or $f(t)=-f(-t)$ (Figure 3.9b), the properties of F will be as summarized in Table 3.4. These properties follow from the cosine and sine

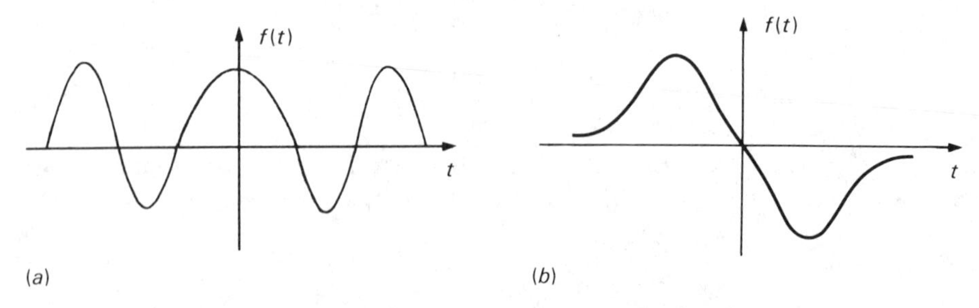

Figure 3.9 Signals of even and odd parity.

Table 3.4 Fourier transformation of a real signal.

$f(t)$	even	odd
$F(\omega)$	real, even	imaginary, odd

being, respectively, even and odd, and from the fact that the integral from $-t$ to t of an odd function is 0.

Any signal f may be split (Figure 3.10) into the sum of an even part f_+ and an odd part f_-:

$$f(t) = \underbrace{\tfrac{1}{2}(f(t) + f(-t))}_{f_+(t)} + \underbrace{\tfrac{1}{2}(f(t) - f(-t))}_{f_-(t)}$$

which sometimes may facilitate the practical evaluation of Fourier transforms and related integrals. This application is illustrated in Example 3.14.

EXAMPLE 3.14

Find the integral $\int_0^\infty e^{-t}\cos(\omega t)\,dt$.

Solution. The integral must be half of

$$\int_{-\infty}^{\infty} e^{-|t|}\cos(\omega t)\,dt = \int_{-\infty}^{\infty} e^{-|t|}e^{-i\omega t}\,dt$$

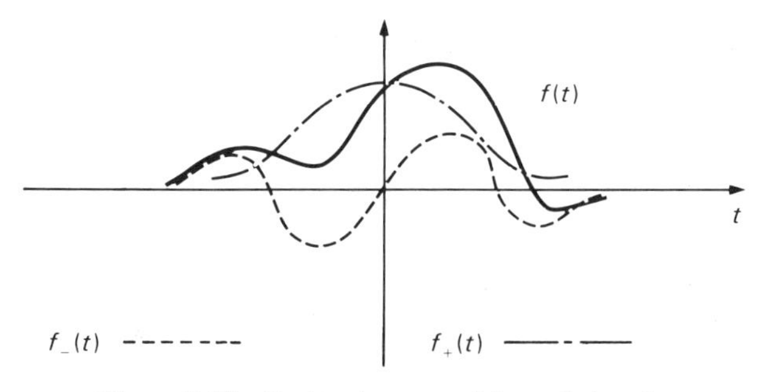

Figure 3.10 Parity decomposition of signal.

which, apart from the factor $(2\pi)^{-1/2}$, is the Fourier transform $F(\omega)$ of the signal $f(t) = e^{-|t|}$, easily found to be

$$F(\omega) = \frac{1}{\sqrt{2\pi}} \frac{2}{(1+\omega^2)}$$

Thus, the integral sought is $1/(1+\omega^2)$.

From Table 3.3, the following relation is seen to hold for arbitrary complex signals f

$$f(t)^* = \frac{1}{\sqrt{2\pi}} \int_{-\infty}^{\infty} F(\omega)^* e^{-i\omega t} \, d\omega \tag{3.25}$$

which may be interpreted as saying that one will get back to f from F via the sequence conjugation–Fourier transformation–conjugation. If, in particular, f is real, the *inverse* Fourier transformation consists in a conjugation followed by a *direct* Fourier transformation. If, finally, we restrict ourselves to real even signals, the inverse and the direct Fourier transforms are identical.

EXAMPLE 3.15

Find the Fourier transform of the signal

$$f(t) = 1/(1+t^2)$$

Finally, we give the analog versions of equations (3.13) and (3.14), that is,

$$\int_{-\infty}^{\infty} f(t) \, dt = \sqrt{2\pi} F(0) \qquad \text{and} \qquad \int_{-\infty}^{\infty} F(\omega) \, d\omega = \sqrt{2\pi} f(0) \tag{3.26}$$

$$f \cdot g = F \cdot G \tag{3.27}$$

where, once again, F and G denote the Fourier transforms of, respectively, f and g, while $F \cdot G = \int F(\omega) G(\omega)^* \, d\omega$ is their scalar product. From equation (3.27) follows the important special case

$$\int_{-\infty}^{\infty} |f(t)|^2 \, dt = \int_{-\infty}^{\infty} |F(\omega)|^2 \, d\omega \tag{3.28}$$

As for equations (3.26), these follow from Table 3.3 on substituting $\omega = 0$ and $t = 0$, respectively, and equation (3.27) follows from the corresponding digital version (3.14), divided by N on both sides of the equality. Next, equation (2.22) is brought into play on both sides, and finally we choose $\Omega = T$ in the derivation on p. 85.

The reader is reminded that $|F(\omega)|$ is called the amplitude spectrum for f. Correspondingly, $|F(\omega)|^2$ is called the *energy spectrum* for f.

EXAMPLE 3.16
Show that

(a) $$\int_{-\infty}^{\infty} \operatorname{sinc}(t)\, dt = \int_{-\infty}^{\infty} \operatorname{sinc}^2(t)\, dt = \pi$$

(b) $$\int_{-\infty}^{\infty} \frac{dt}{1+t^2} = \pi$$

(c) $$\int_{-\infty}^{\infty} \frac{dt}{(1+t^2)^2} = \frac{\pi}{2}$$

using the examples of Fourier-transformed signals given.

3.3.3 The fast Fourier transform

If one has to compute the two-dimensional DFT for an image, the demands upon speed and capacity can easily become prohibitive – unless the process is arranged in a rational manner. True, the DFT for an image of dimensions $M \times N$ may be split into M DFTs of order N, followed by N DFTs of order M. If, however, the expression (3.12) for the one-dimensional DFT is employed, one is obliged to perform a large number of superfluous computations. The steady increase in practical applications of signal analysis is largely due to the simplifications which can be employed. Several algorithms exist for reducing considerably the number of computational operations. They all answer to the name of *fast Fourier transforms* (FFTs), and we now describe a typical specimen of this kind.

In this method, a DFT of order N is broken down into two DFTs each of order $N/2$ (where N is even, say $N = 2M$):

$$\begin{aligned}\sqrt{N}F[m] &= \sum_{n=0}^{N-1} f[n]\, w_N^{-mn} \\ &= \sum_{k=0}^{M-1} f[2k]\, w_N^{-m \cdot 2k} + \sum_{k=0}^{M-1} f[2k+1]\, w_N^{-m \cdot (2k+1)}\end{aligned}$$

The original sum is now divided by taking alternate terms, that is, terms of even and odd indices, respectively. In the latter summation, w_N^{-m} is taken outside, and we note that $w_N^2 = w_M$:

$$\sqrt{N}F[m] = \sum_{k=0}^{M-1} f[2k]\, w_M^{-mk} + w_{2M}^{-m} \sum_{k=0}^{M-1} f[2k+1]\, w_M^{-mk}$$

Both sums are formally DFTs of order $M = N/2$ and have been built from the signal values of even (odd) digital time. They *are* DFTs, if $m \in [0, M-1]$. If, on the other hand, $m \in [M, 2M-1]$, we may write $m = M + m'$ with $m' \in [0, M-1]$:

$$\sqrt{N}F[m] = \sum_{k=0}^{M-1} f[2k]\, w_M^{-(M+m')k} + w_{2M}^{-(M+m')} \sum_{k=0}^{M-1} f[2k+1]\, w_M^{-(M+m')k}$$

and, as $w_M^{-M} = 1$ and $w_{2M}^{-M} = -1$, we further obtain

$$\sqrt{N}F[m] = \sum_{k=0}^{M-1} f[2k]\, w_M^{-m'k} - w_{2M}^{-m'} \sum_{k=0}^{M-1} f[2k+1]\, w_M^{-m'k}$$

To summarize, for $m \in [0, M-1]$:

$$F[m] = (F_{\text{even}}[m] + w_{2M}^{-m} F_{\text{odd}}[m])/\sqrt{2} \tag{3.29}$$

$$F[M+m] = (F_{\text{even}}[m] - w_{2M}^{-m} F_{\text{odd}}[m])/\sqrt{2} \tag{3.30}$$

Here, F is a DFT of order $N = 2M$, while F_{even} and F_{odd} are DFTs of order M. If M itself is even, the process may be continued, and if N is a power of 2, the process may be reduced to a DFT of order 2.

An enumeration of the number of *multiplications* appearing in the FFT algorithm developed here provides a glimpse of the method's power. First of all, we note that a computation of the DFT according to equation (3.12) requires $N-1$ multiplications for each $m \in [1, N-1]$, a total of $(N-1)^2$ multiplications (we disregard the N *distinct* values w_N^{-mn} which are assumed to be computed once and for all and stored for subsequent use).

Next, equations (3.29)–(3.30) show that if a DFT of order M requires a multiplications, then $2a + M$ multiplications will be needed for order $2M$. If the algorithm can be broken down to order 2, that is, if $M = 2^p$, we can compute the number of multiplications by starting with $p = 1$ and using this relation successively. Some of the results are collected in Table 3.5. In Section 3.7 it will be shown that the number is given by

$$a[p] = 2^{p-1}p = \tfrac{1}{2} N \log_2 N$$

where $\log_2$ denotes the logarithm of base 2.

EXAMPLE 3.17

Prove the above formula by mathematical induction. Then show that the corresponding number of additions is $2^p p = N \log_2 N$.

EXAMPLE 3.18

Find the number of multiplications when performing an FFT on an image having 1024×512 pixels. If the image is separable, what does this number become?

Table 3.5 Computational reductions using fast Fourier transforms.

p	$N=2^p$	$(N-1)^2$	$a[p]$	Reduction factor
1	2	1	1	1.00
2	4	9	4	2.25
3	8	49	12	4.08
4	16	225	32	7.03
5	32	961	80	12.01
6	64	3 969	192	20.67
7	128	16 129	448	36.00
8	256	65 025	1024	63.50
9	512	261 121	2304	113.33
10	1024	1 046 529	5120	204.40

The practical implementation of the FFT algorithm may take place in various ways. Appendix B gives a BASIC subroutine performing the transformation according to the principles outlined here.

3.4 Systems

When the word 'system' is encountered in the context of images, one is led either to think of material objects designed to store, transmit, or process data ('hardware'), or to ponder program packages implemented on a computer to solve certain image processing tasks ('software').

In the image-processing literature, however, a third kind of usage is encountered. *Systems* are special transformations, as described in the previous section, possessing the additional property of *time invariance*. This means, very loosely stated, that the transformation produces its output regardless of the time at which it is presented with its input (cf. Figure 3.11).

For images, where 'time' is to be replaced by 'position', one and the same object is transformed in the same way by the system, no matter where the transformation takes place in the image.

A transformation S is called *(time-)invariant* if – for all possible input signals f and all complex numbers k – it obeys

$$S(f[n-k]) = Sf[n-k] \tag{3.31}$$

On the left-hand side, $f[n-k]$ represents the signal resulting from applying a time-translation by k units to f, that is, $T_k f$. The right-hand side gives the result of applying the same translation to the 'original' output Sf.

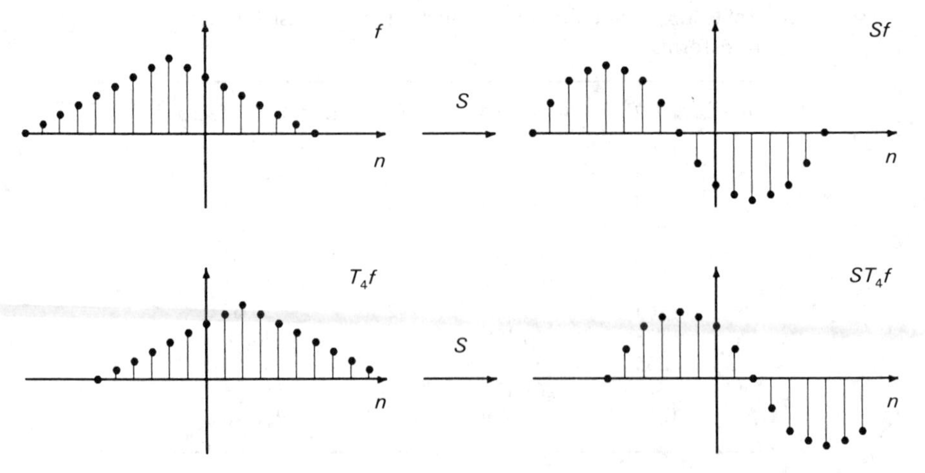

Figure 3.11 Effect of an invariant transformation (system).

Expressed in words: an arbitrary shift of the input results in the same shift of the output signal. Accordingly, S is invariant if and only if

$$ST_k = T_k S$$

that is, if S commutes with all the translations T.

EXAMPLE 3.19
Explain that it is sufficient to assume $ST_1 = T_1 S$ to obtain invariance.

Such an invariant transformation is termed a *system*. As will be demonstrated in the remainder of this chapter, this seemingly very innocuous property opens the gate to a flood of aspects, the subject of (linear) *systems theory*, a glimpse of which will now be provided.

For images, the definition of 'system' is as follows:

$$S(b[m-k, n-l]) = Sb[m-k, n-l] \tag{3.32}$$

or, using an even more compact notation (cf. Example 3.19),

$$ST_{10} = T_{10}S \qquad \text{and} \qquad ST_{01} = T_{01}S$$

EXAMPLE 3.20
The following transformations are *not* systems:

(a) $Tf[n] = nf[n]$ (why not?)

(b) A rotation of an image about, for example, its centre. An object near the edge of the image will be moved through a larger distance than a similar object near the image centre.

3.5 Characterization of systems

The fundamental property of systems allows a complete characterization of the effect of a system in extremely simple terms. In what follows, we shall focus on two complementary descriptions. The first draws on a knowledge of the system's *impulse response*, as was done earlier (cf. Section 3.2); the second concentrates on the *exponential response*, the response produced by an exponential (or harmonic) input.

3.5.1 The impulse response

For a given system S, we denote its δ- or impulse response by h:

$$h[n] = S\delta[n] \tag{3.33}$$

As shown by the following considerations, S is completely characterized by $h[n]$.

If, once again, $h_k[n]$ stands for $S(\delta[n-k])$, the invariance gives

$$h_k[n] = h[n-k]$$

after which the system's action on a signal $f[n]$, according to equation (3.9), is given by

$$Sf[n] = \sum_k f[k]\,h[n-k] \tag{3.34}$$

The above operation producing $g[n] = Sf[n]$ from $f[n]$ and $h[n]$ is called a (digital) *convolution* and is represented by the symbol $\star$:

$$g = f \star h \tag{3.35}$$

The convolution of two signals f_1 and f_2 is thus defined by

$$f_1 \star f_2[n] = \sum_m f_1[m]\,f_2[n-m] \tag{3.36}$$

which also may be written

$$f_1 \star f_2[n] = \sum_{k+l=n} f_1[k]\,f_2[l]$$

It follows that

$$f_1 \star f_2[n] = f_2 \star f_1[n] \tag{3.37}$$

This leads to the following important result:

> The output of a system is equal to the convolution of the input with the impulse response of the system.

This characterizes the action of the system completely in terms of $h[n]$.

EXAMPLE 3.21
Show that

$$\sum_n f_1 \star f_2[n] = \left(\sum_n f_1[n]\right)\left(\sum_n f_2[n]\right)$$

EXAMPLE 3.22
If a system S has an impulse response h given by

$$h[n] = \delta[n-1] + 2\delta[n] + \delta[n+1]$$

(Figure 3.12a), the response g to the input $f[n]$ is

$$g[n] = \sum_k f[k]h[n-k] = f[n-1] + 2f[n] + f[n+1]$$

For example, the input

$$f[n] = \delta[n] + 2\delta[n-1] + 3\delta[n-2]$$

(Figure 3.12b) yields

$$g[n] = \delta[n+1] + 4\delta[n] + 8\delta[n-1] + 8\delta[n-2] + 3\delta[n-3]$$

where $g[1] = f[0] + 2f[1] + f[2] = 1 + 2 \cdot 2 + 3 = 8$, etc. (Figure 3.12c).

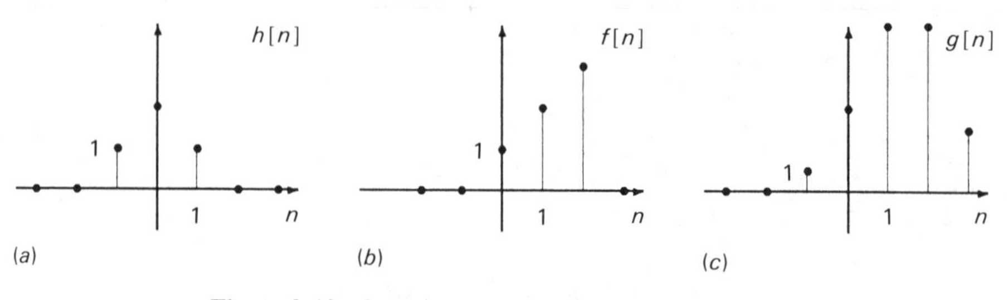

Figure 3.12 Impulse response, input, and output.

What is the response g to

(a) $f[n] = h[n-1]$

(b) $f[n] = r_4[n+2]$

and what are $g[0]$, $g[1]$, and $g[2]$ in the following cases:

(c) $f[n] = s[n]$

(d) $f[n] = 2^n$?

The composite of two systems S_1 and S_2 having impulse responses h_1 and h_2, respectively, is another system $S = S_2 S_1$ of impulse response h:

$$h = S\delta = S_2(S_1\delta) = S_2 h_1 = h_2 \star h_1 \tag{3.38}$$

The impulse response of the composite system is thus equal to the convolution of the individual impulse responses. From this, we derive the important rule that

Systems are always commutative.

EXAMPLE 3.23
The systems S_1 and S_2 have impulse responses

$$h_1[n] = \tfrac{1}{2} r_2[n] \qquad \text{and} \qquad h_2[n] = \tfrac{1}{4} r_4[n]$$

Find the impulse responses for the three systems $S_1^2 = S_1 S_1$, $S_2^2 = S_2 S_2$, and $S_1 S_2$.

For images, the impulse response is itself an image specifying the system's effect upon a point image or a point source. The image impulse response is traditionally called the *point spread function* (PSF):

$$PSF[m,n] = S\delta[m,n] \tag{3.39}$$

and the system's action of changing an image b into $c = Sb$ thus consists in

$$c[m,n] = \sum_k \sum_l b[k,l]\, PSF[m-k, n-l] \tag{3.40}$$

EXAMPLE 3.24
An image-processing system is characterized by the point spread function

$$PSF[m,n] = 0.6\delta[m,n] + 0.1\delta[m-1,n] + 0.1\delta[m,n-1] + 0.1\delta[m+1,n] + 0.1\delta[m,n+1]$$

0	0	0	0	0
0	0	0.1	0	0
0	0.1	0.6	0.1	0
0	0	0.1	0	0
0	0	0	0	0

as shown in the figure above. The system processes an image, a portion of which is shown below:

247	143	351	182	182	65
	110	270 (shaded)	140	(shaded)	
	165	405			

What is the value, after processing, in the shaded pixel within the portion shown?

The portion is next specified to be part of a separable image. What is the value in the shaded pixel outside, both before and after processing?

3.5.2 The exponential response

As mentioned earlier, systems possess one extremely important characteristic property: they amplify exponential signals. To see how, let S be a system with input

$$f[n] = z^n$$

(where, for harmonic signals, $|z| = 1$). If the response Sf is called g, linearity and invariance show that

$$\begin{aligned} g[n+m] &= Sf[n+m] = S(z^{n+m}) = z^m S(z^n) \\ &= z^m g[n] \end{aligned} \tag{3.41}$$

for all values of the translation parameter m. For digital time $n = 0$, we have $g[m] = z^m g[0]$. But, as this result holds for all m, the response g is completely specified:

$$g[n] = g[0]\, z^n$$

that is to say:

> The effect of an arbitrary system upon an exponential signal is multiplication by a constant factor.

The factor $g[0] = Sf[0]$ depends only on the base z for the exponential signal; accordingly, it may be written as $H(z)$:

$$S(z^n) = H(z)z^n \tag{3.42}$$

The quantity $H(z)$ is called the *system function*. In the next section, we turn to its determination in practice.

EXAMPLE 3.25
Find the system function for

(a) the system in Example 3.22;
(b) the system with impulse response $h[n] = 3^{-n}s[n]$.

Thus if a signal f has been expressed as a linear combination of exponential signals, the action of the system upon f consists in an amplification (multiplication by a constant number) of each component. The resulting signal is, therefore, a *new* linear combination of the *same* exponential signals; hence, it is completely determined from the knowledge of the system function.

3.6 The z-transformation

The distinctive property of systems, as compared to other transformations, is the fact that a complete description is contained in a single signal, the impulse response. This, in a certain sense, puts 'signals' and 'systems' on an equal footing – for instance, the principles of information reduction may, without further ado, be taken over from signals to systems. Moreover, the action of the system, as expressed in equation (3.34), may be directly reformulated; in place of a convolution with the impulse response, a *multiplication by the system function* appears – greatly facilitating practical computation. This is accomplished by means of the so-called *z-transformation*, which will be introduced in the following.

Once again, we send the exponential signal

$$f[n] = z^n$$

through the system S, characterized by the impulse response $h[n]$. This leads to an output

$$\begin{aligned} g[n] &= \sum_m f[n-m]\,h[m] = \sum z^{n-m}h[m] \\ &= \left(\sum z^{-m}h[m]\right)z^n \end{aligned}$$

As shown above, the result consists in an amplification of the signal f. The multiplication factor, called the system function, is thus given by

$$H(z) = \sum_{m=-\infty}^{\infty} h[m]\,z^{-m} \tag{3.43}$$

The system function $H(z)$ may therefore be expressed as a power series in z, the coefficients of which are the values $h[m]$ of the impulse response. Such a power series is called a *z-transform*. The term z-transformation is the traditional one, but it is nevertheless somewhat misleading in view of the usage in the present chapter. The z-transformation is not a signal transformation, as it replaces a digital signal $f[n]$ with a continuous function $F(z) = \Sigma f[n] z^{-n}$. Nevertheless, there is a close connection. If the signal $f[n]$ differs from zero only for the digital times $n \in [0, N-1]$, and its z-transform is $F(z)$, the DFT for f consists of the N values $F(w_N^m)/\sqrt{N}$, $m \in [0, N-1]$. Thus, the DFT consists (apart from the factor $\sqrt{N}$) of N samples of the z-transform.

Defining the z-transform $F(z)$ of a signal $f[n]$ by

$$F(z) = \sum_{n=-\infty}^{\infty} f[n] z^{-n} \tag{3.44}$$

the result (3.43) may be formulated:

The system function is the z-transform of the impulse response.

EXAMPLE 3.26

The z-transform of the rectangular signal $r_N[n]$ is

$$R_N(z) = \sum_{n=0}^{N-1} z^{-n} = \frac{1 - z^{-N}}{1 - z^{-1}}$$

Consequently, the z-transform of the step signal $s[n]$ is

$$S(z) = \frac{1}{1 - z^{-1}} = \frac{z}{z - 1}$$

Images are z-transformed in an analogous way:

$$B(u, v) = \sum_m \sum_n b[m, n] u^{-m} v^{-n}$$

Here, $B(u, v)$ denotes the two-dimensional z-transform of the image b. If b is separable, that is, if $b[m, n] = f[m]\, g[n]$, then

$$\begin{aligned} B(u, v) &= \sum_m \sum_n f[m]\, g[n]\, u^{-m} v^{-n} = \sum_m f[m]\, u^{-m} \sum_n g[n]\, v^{-n} \\ &= F(u)G(v) \end{aligned}$$

where F and G are the (one-dimensional) z-transforms of f and g.

EXAMPLE 3.27
Find the z-transforms of the images

(a) $b[m, n] = 3^{-m} s[m] s[n]$

(b) $b[m, n] = s[m] \delta[n - 2m + 3]$

The evaluation of a z-transform thus consists in the *summation of a power series* in the variable z. The result is, in principle, a continuous function of z. In practice, only a finite number of function values will be accessible.

When performing the *inverse* z-transformation, the task is a determination of all the coefficients in the power series of a given function. In equation (3.44), the function F is given, and the coefficients $f[n]$ are to be determined. Thus, if there are N unknown coefficients, it should be possible to solve for these by selecting N values of z, writing the relation (3.44) for these values and solving the resulting system of equations.

The advantage of the z-transformation lies with the following important result:

> The z-transform of a system output equals the product of the system function and the z-transform of the input.

This is shown as follows. If a system S, with impulse response $h[n]$, is fed the input $f[n]$ and produces the output $g[n]$, then

$$g[n] = \sum_m f[n-m] h[m] \tag{3.45}$$

This equation is now z-transformed:

$$\begin{aligned}\sum_n g[n] z^{-n} &= \sum_n \sum_m f[n-m] h[m] z^{-n} \\ &= \sum_n \sum_m f[n-m] z^{-(n-m)} h[m] z^{-m} \\ &= \sum_n f[n] z^{-n} \sum_m h[m] z^{-m}\end{aligned}$$

The two single sums appearing after the last equality sign are the z-transforms of f and h, respectively. If all transforms are denoted by capital letters, we have thus obtained the following result:

$$G(z) = F(z) H(z) \tag{3.46}$$

Since $H(z)$ is the system function, we have proved our assertion above.

Since a sum of products like that in equation (3.45) is called a convolution, the above result may also be stated as follows:

> The z-transform of a convolution of two signals equals the product of the two individual z-transforms.

In particular, the system function for a composite system is the product of the individual system functions (cf. equation (3.38)).

3.6.1 The inverse z-transformation

Among the z-transforms encountered in theoretical considerations, many are rational functions in z (or z^{-1}), that is, of type $R(z) = P(z)/Q(z)$, P and Q being polynomials. Rational functions may, by various means, be rewritten as sums of other rational functions, all of form like the polynomial in the left column of Table 3.6. Here, p and q are positive integers, a a real number not equal to 0. In order to determine the inverse z-transform of $R(z)$, it suffices to do so for the simple expression in Table 3.6.

To prove the result in Table 3.6, we draw upon three other useful results:

1. (Shifting.) If $F(z)$ is the z-transform of $f[n]$, the z-transform of $f[n+k]$ is equal to $z^k F(z)$:

$$\begin{aligned}\sum_{n=-\infty}^{\infty} f[n+k]z^{-n} &= z^k \sum_{n=-\infty}^{\infty} f[n+k]z^{-(n+k)}\\ &= z^k \sum_{n+k=-\infty}^{\infty} f[n+k]z^{-(n+k)}\\ &= z^k F(z)\end{aligned} \tag{3.47}$$

2. (z-transform of an exponential signal.) If $f[n] = a^n s[n]$, then $F(z) = z/(z-a)$ (for $|z| > a$):

$$F(z) = \sum_{n=0}^{\infty} a^n z^{-n} = \frac{1}{1-(a/z)} = \frac{z}{z-a} \tag{3.48}$$

Table 3.6 Inverse z-transform of a rational function.

z-transform	Signal
$A\dfrac{z^p}{(z-a)^q}$	$A\dbinom{n+p-1}{q-1}a^{n+p-q}s[n+p-q]$

3. (Differentiation.) In equation (3.48), we take the mth derivative with respect to a:

$$\sum_{n=m}^{\infty} n(n-1)\cdots(n-m+1)a^{n-m}z^{-n} = m!\,\frac{z}{(z-a)^{m+1}} \tag{3.49}$$

This expresses the fact that the z-transform of $f[n] = \binom{n}{m}a^{n-m}s[n-m]$ equals $z/(z-a)^{m+1}$. On combining results 1 and 3, we finally obtain the result stated in Table 3.6.

EXAMPLE 3.28

Find the inverse z-transform of

(a) $F(z) = \dfrac{3z^2 - 2z + 1}{z^3 - 2z^2 + z}$ (b) $F(z) = \dfrac{3z^3 + 13z^2 + 14z}{z^2 + 4z + 3}$

EXAMPLE 3.29

We wish to evaluate the sum of all squares from 1 to N. This sum can be written as a convolution:

$$f[n] = (n^2 s[n]) \star s[n]$$

where the sum in question is $f[N]$. Since $n^2 = \binom{n}{2} + \binom{n}{1}$, z-transformation of the above equation yields

$$F(z) = \left(2\,\frac{z}{(z-1)^3} + \frac{z}{(z-1)^2}\right)\frac{z}{z-1}$$

$$= \frac{z^3}{(z-1)^4} + \frac{z^2}{(z-1)^4}$$

The inverse z-transform is

$$f[n] = \binom{n+2}{3}s[n-1] + \binom{n+1}{3}s[n-2]$$

and, for $n = N$,

$$f[N] = \frac{(N+2)(N+1)N}{6} + \frac{(N+1)N(N-1)}{6}$$

so that

$$\sum_{n=1}^{N} n^2 = \tfrac{1}{6}N(N+1)(2N+1)$$

EXAMPLE 3.30

Carry out the summation

$$\sum_{n=1}^{N-1} n^2(N-n)^2$$

(*Hint*: The sum is a convolution of $n^2 s[n]$ with itself.)

EXAMPLE 3.31

In order to calculate the effect of a system S, given by its δ-response h, upon an arbitrary input f, it is possible (according to equation (3.46)) to replace the direct convolution (equation (3.34)) by the seemingly more complicated process involving a z-transformation of f and h, multiplication of the results, and finally, an inverse z-transformation of this product. This procedure, however, normally leads to enormous reductions in computing time, as exemplified in the following.

A system S is described by an impulse response h, in which $h[n] = 0$ for $n < 0$ and $n \geqslant 20$. The system acts upon an input f for which $f[n] = 0$ for $n < 0$ and $n \geqslant 1000$. The output signal thus consists of at most 1019 non-zero values. If these values are calculated directly from equation (3.34), we need about 20 000 multiplications per output value, which is of the order of 20 million operations.

The second method requires, first, two z-transformations. It appears reasonable to expect that a knowledge of these at only 1019 z-points will suffice for a reconstruction of the 1019 output values. To this end, we attempt to estimate how many multiplications are required.

For a given z, we first compute the values $z^{-1}, z^{-2}, \ldots, z^{-1019}$. These are found by successive multiplication by z^{-1} – that is, 1018 multiplications in all. These values are stored for later use. Next, we calculate $F(z) = \Sigma\, f[n] z^{-n}$ (1000 multiplications), $H(z) = \Sigma\, h[n] z^{-n}$ (20 multiplications), and $G(z) = F(z) \cdot H(z)$ (1 multiplication). To compute the specific value $G(z)$, we have carried out 2039 multiplications, and we must calculate 1019 of these. Thus, the total number of multiplications involved in the determination of G will be approximately 2 million.

If the z-transformation is performed at the points $z_k = w_N^k$, that is, as for the DFT, the inverse transformation of G can be carried out according to the precept outlined earlier, and the number of multiplications is the same as that occurring in a direct DFT, slightly above 1 million. The total number of multiplications in this line of attack upon $g[n]$ is thus only about 3 million. The computing time in the seemingly roundabout method will thus – all other things being equal – amount to only some 15 per cent of that of the direct method.

If, finally, the process is carried out as an FFT of order 1024, Table 3.5 shows that each of the three Fourier transformations requires some 5000 multiplications only (and, because of the many zeros, the computation of H will proceed considerably more rapidly). The total computing time is thus reduced by a factor of at least 3000 compared to that taken by the original 20 million multiplications.

3.7 Examples of systems

3.7.1 Difference equations

An equation such as the following:

$$g[n+1] - 2g[n] = f[n] \tag{3.50}$$

is called a *difference equation*. The 'unknown' is the signal $g[n]$; $f[n]$ is known. Since equation (3.50) is linear and time-invariant – that is, identical to the equation resulting from an arbitrary time translation – it seems worth investigating whether the change from f to g defines a system.

Clearly, if only one value of $g[n]$ is known, all other values of g may be found successively. If, for instance, $f[n] = \delta[n]$ and $g[0] = 1$ is specified, then

$$g[1] = f[0] + 2g[0] = 3,\ g[2] = f[1] + 2g[1] = 6, \ldots$$

and

$$g[-1] = \tfrac{1}{2}(g[0] - f[-1]) = \tfrac{1}{2},\ g[-2] = \tfrac{1}{2}(g[-1] - f[-2]) = \tfrac{1}{4}, \ldots$$

The given value, in this case $g[0]$, is called the *initial value*. Note that 0 is the only initial value compatible with linearity between f and g.

Due partly to the ambiguity thus introduced, and partly to the specific time (conflicting with time invariance), the difference equation does not define a system right away. The remedy is to fix one of the values in the impulse response h. In this example, one would often choose $h[0] = 0$, leading to $h[n] = 0$ for all $n < 0$. The system defined in this way, where each input value produces an output value at a *later* time, is termed *causal*.

With this choice, h is given by

$$h[n] = 2^{n-1}s[n-1]$$

The system function H is the z-transform of this signal; according to Table 3.6,

$$H(z) = z^{-1}\frac{z}{z-2} = \frac{1}{z-2} \tag{3.51}$$

as is seen by z-transforming both sides of equation (3.50).

In general, a difference equation 'with constant coefficients' is of the form

$$\sum_k a_k g[n+k] = \sum_l b_l f[n+l]$$

The difference equation defines a system if a number of impulse response values are specified. Extending the above example, we find that the system function is

$$H(z) = \sum_l b_l z^{n+l} \Big/ \sum_k a_k z^{n+k} \tag{3.52}$$

EXAMPLE 3.32

The number of multiplications in the FFT algorithm obeys equation (3.50) – cf. p. 88 – if the input is taken to be

$$f[n] = 2^n s[n]$$

This signal has a z-transform equal to

$$F(z) = \frac{z}{z-2}$$

from which it follows that

$$A(z) = F(z)H(z) = \frac{z}{(z-2)^2}$$

which has inverse z-transform (see Table 3.6)

$$a[n] = \binom{n}{1} 2^{n+1-2} s[n-1]$$

so that $a[p] = 2^{p-1}p$, as stated earlier.

3.7.2 Mechanical systems

The fundamental equation for mechanical systems is Newton's second law, according to which the force f upon a particle is the product of the particle's mass and its acceleration. If the particle moves in one dimension only, as specified by the coordinate x, its velocity is $v = \mathrm{d}x/\mathrm{d}t$ and the acceleration $a = \mathrm{d}^2x/\mathrm{d}t^2$. If the mass of the particle is m, the equation of motion says

$$ma = f$$

The relation between force $f(t)$ and position $x(t)$ defines a system having $f(t)$ as the arbitrary input and $x(t)$ as output – the force f causes a particle position x – and the initial position and velocity are assumed to be specified in a time-invariant manner.

A classical mechanical problem is the 'forced harmonic oscillator'. Here, the input force $f = A\mathrm{e}^{\mathrm{i}\omega t}$ appears as part of the total force

$$f + f_1 + f_2 = A\mathrm{e}^{\mathrm{i}\omega t} - \gamma v - kx$$

The last component, $f_2 = -kx$, keeps the particle near an equilibrium position $x = 0$; any deviations will result in a restoring force towards this position. A simple spring provides a reasonable approximation to this behaviour.

The second component, $f_1 = -\gamma v$, represents a frictional force proportional to the velocity v and of opposite direction. Finally, $f = Ae^{i\omega t}$ is the 'external' force, varying harmonically with time, and regarded as the 'proper' input.

The relation between f and x is now given by the differential equation

$$m\frac{d^2x}{dt^2} + \gamma\frac{dx}{dt} + kx = Ae^{i\omega t} \tag{3.53}$$

which, then, defines a system. The response to the harmonic input $f = Ae^{i\omega t}$ is

$$x(t) = H(e^{i\omega})f(t) = H(e^{i\omega})Ae^{i\omega t}$$

which should obey

$$mH(e^{i\omega})(i\omega)^2Ae^{i\omega t} + \gamma H(e^{i\omega})i\omega Ae^{i\omega t} + kH(e^{i\omega})Ae^{i\omega t} = Ae^{i\omega t}$$

that is,

$$H(e^{i\omega}) = \frac{1}{k - m\omega^2 + i\gamma\omega} \tag{3.54}$$

Defining $a = k - m\omega^2$ and $b = \gamma\omega$, then $H(e^{i\omega}) = 1/(a + ib) = (a - ib)/(a^2 + b^2)$, and with $H(e^{i\omega}) = |H(e^{i\omega})|e^{i\phi}$, we have

$$|H(e^{i\omega})| = \frac{1}{\sqrt{a^2 + b^2}} = \frac{1}{\sqrt{(k - m\omega^2)^2 + \gamma^2\omega^2}}$$

$$\tan\phi = -\frac{b}{a} = \frac{-\gamma\omega}{k - m\omega^2}$$

Since the system effect quite generally may be factored into an *amplification* by $|H(e^{i\omega})|$ and a *phase shift* ϕ, these two quantities are given by the above equations in the case of the forced harmonic oscillator.

Due to the large formal similarities between mechanical and electronic systems – both are described mathematically by differential equations of second order (cf. equation (3.53)) – the same expressions for amplification and phase shift appear, for example, for alternating-current circuits.

3.7.3 Electronic circuits

The classical so-called passive elements in an electronic circuit are the resistor, capacitor, and inductor. Each of these may be regarded as a system with the external voltage V as input and the resulting electric current I through the circuit as output.

For a circuit built from many such components, the task will often be to evaluate the current through a fixed point in the circuit, given an externally imposed voltage difference between two points in it. To accomplish this, one uses Kirchhoff's laws. These laws are linear and time-invariant; thus, the classical circuit elements define a system.

Quite often, the external voltage is a harmonically alternating one:

$$V = V_0 e^{i\omega t}$$

This voltage results in a current $I(t)$, which is found from $V(t)$ by multiplication with the system function:

$$I = H(e^{i\omega})V$$

The traditional term for the quantity $Z(\omega) = 1/H(e^{i\omega})$ is the *(generalized) impedance*, so that an alternating current of frequency ω will lead to

$$V = Z(\omega)I$$

For the system consisting of a single resistor R only, $Z(\omega) = R$.

3.7.4 Electromagnetic waves and aperture distributions

The propagation of electromagnetic waves may also be described in terms of systems. We adopt the point of view that input and output, respectively, correspond to the electric field at two points along the direction of propagation, say at the points 0 and x. The field at the point 0 is thus considered as causing the field at the point x. The relevant system is the 'empty space' between the points.

The complete variation of the field is, according to equation (1.9),

$$E(x, t) = Ae^{i(\omega t - kx)}$$

If the input signal – that is, $E(0, t)$ – equals $e^{i\omega t}$, then $E(x, t) = e^{i(\omega t - kx)} = e^{-ikx}e^{i\omega t}$. The system effect is thus a multiplication by $e^{-ikx} = e^{-i\omega x/c}$, which fixes the system function on the complex unit circle:

$$H(e^{i\omega}) = e^{-i\omega x/c} \tag{3.55}$$

From this result, we can determine the spatial variation of the field across a receiver area – for instance, the aperture of an optical telescope or a radio telescope. In Figure 3.13, the radiation source is placed somewhere on the far left, and the telescope aperture is located on the right. The position in the plane aperture is specified by an (x, y)-coordinate system. The directions to the points in the distant source are given by a unit vector e which itself is specified (unambiguously) by means of its projection (x, y) on to the aperture plane. The spatial distribution of amplitude across the source may then be regarded as an image $A = A(x, y)$.

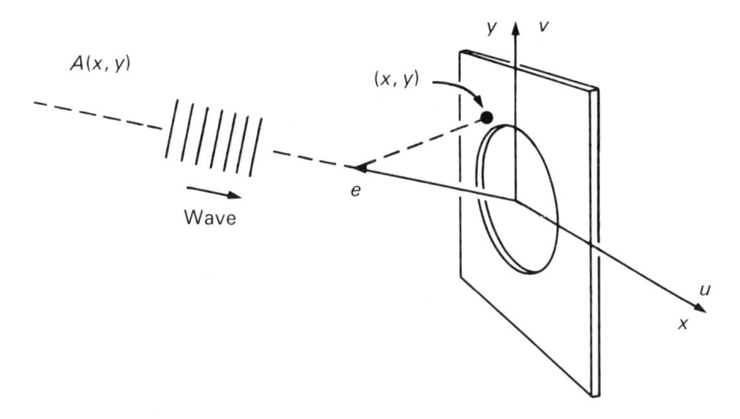

Figure 3.13 Geometry for source and aperture.

The radiation arriving along parallel paths from the source point (x, y) must cover different distances in order to reach the different points in the aperture plane. The difference in path length for arrivals in, respectively, $(0, 0)$ and (u, v), is

$$e \cdot (u, v) = xu + yv$$

and the field distribution across the aperture will accordingly be

$$a(u, v) = \int \int A(x, y) \mathrm{e}^{-\mathrm{i}\omega(xu + yv)/c} \, \mathrm{d}x \, \mathrm{d}y \tag{3.56}$$

Apart from the constant $1/c$ – which, if necessary, may be absorbed in the units for (x, y) and/or (u, v) – and the measuring constant appearing in the field and the amplitude, the aperture distribution is obtained by a Fourier transformation of the source distribution.

Thus, when an optical or radio astronomical receiver system produces an image of the radiation source, it performs an (inverse) Fourier transformation of the aperture distribution.

3.7.5 Astronomical images

Figure 3.14 is a CCD image of a distant galaxy (the spot in the middle of the image). The three bright objects are stars in our own galaxy, the Milky Way. The distance to the galaxy is several billion light years, whereas the stars are located 'only' perhaps a thousand light years away. The physical extension of the stars is, in the present context, negligibly small – accordingly, they may be regarded as point sources, that is, impulse (δ-)signals. The reason for the stars' being recorded as extended images is air turbulence in the Earth's atmosphere. This atmospheric

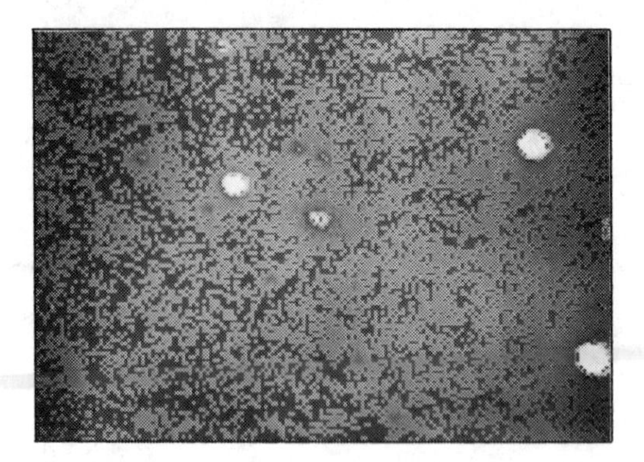

Figure 3.14 A distant elliptical galaxy.

tremor modifies the images of the sources, and we wish to investigate whether this change defines a system S.

If a star has image coordinates $[k, l]$, the intensity values in the image yield precisely $S(\delta[m-k, n-l])$. For this image, the three stellar images prove to be identical – except for total intensity and position. This means that $ST_{kl} = T_{kl}S$, at least for choices of $[k, l]$ corresponding to the coordinates of the three stars, a strong indication that S is invariant, that is, a system. This property can be extremely significant when analyzing the CCD image. Using the methods of Chapter 7, it is possible to remove the atmospheric blur. Here, the stellar images are used to determine the PSF due to this smearing.

The result of this image restoration is shown in Figure 3.15. It was carried out in two steps; first, the image noise was removed (Figure 3.15a), after which the correction for the blurring was applied (Figure 3.15b). One star is shown together with the galaxy.

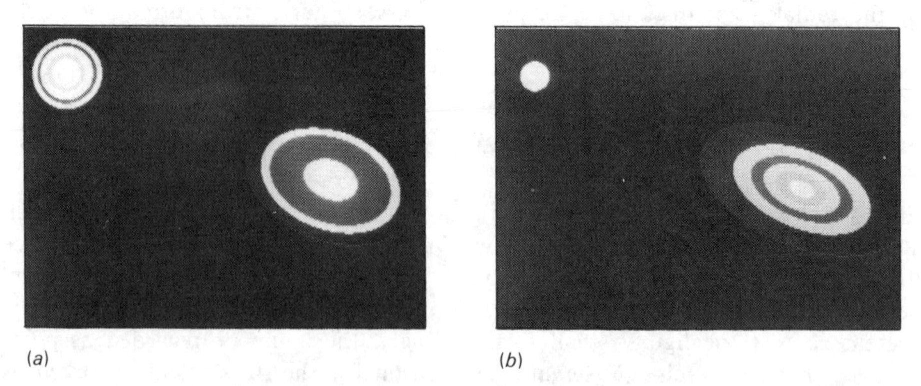

Figure 3.15 Restoration of galaxy image.

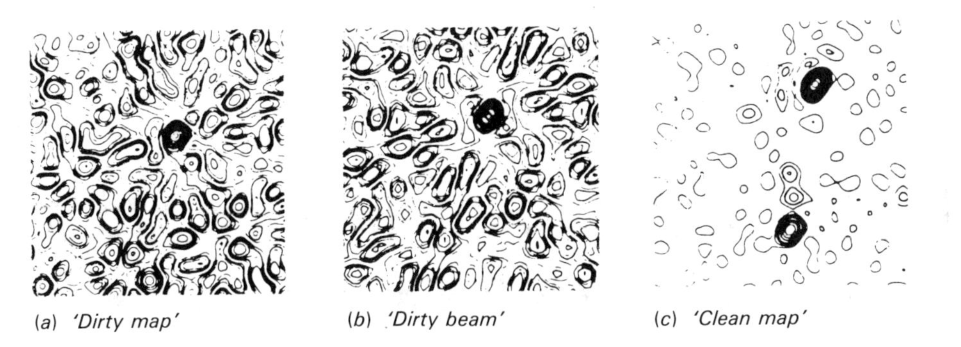

Figure 3.16 Radio PSF placed between observed and 'cleaned' radio map.

The three images in Figure 3.16 illustrate the corresponding situation in radio astronomy. Figure 3.16a is a contour map (an image showing curves between points having the same intensity) of a distant radio source. This image, too, is blurred. In this case, however, the blur is almost exclusively due to the radio telescope itself. The PSF for this blurring mechanism is depicted in Figure 3.16b. Just as above, the knowledge of the PSF enables the construction of the improved contour map as shown in Figure 3.16c. In this figure, the three types of radio maps are distinguished by means of their names from the radio astronomical argot.

4

Coding, Quantization, and Segmentation

A frequently occurring theme in the previous three chapters has been that of information reduction. This feature is intimately linked to the existence of certain types of image structure. The quantization and digitization schemes were, for instance, brought about by image quantities exhibiting 'slow' variation across the image. And the satisfactory reproduction of the image in Figure 2.26 with relatively few harmonic components was, characteristically, caused by the modest content of high-frequency Fourier components.

Very loosely speaking, the image or signal information may generally be split into two components, a 'qualitative' and a 'quantitative' part. The latter component is, by definition, the one remaining after the former has been taken into account.

EXAMPLE 4.1

In the analysis of Figure 2.26 into harmonic components, the image is, in effect, represented by its Fourier coefficients. The 'qualitative' information encompasses the *knowledge* of the significance of these coefficients – that is, the necessity for the user to construct harmonic basis images, multiply by the coefficients in question, and add the resulting images. The 'quantitative' information consists of the coefficients themselves, considered as a set of data with no internal structure. Nevertheless, these coefficients will in general exhibit a marked 'residual' structure, for example, a decrease in absolute value with the order K of the harmonic image representation. It is thus possible to extend the decomposition of the information content, and the reader is urged to pursue this line of attack.

Another obvious example of the distinction between qualitative and quantitative information is provided by Figure 2.4 on p. 34. Here, the quantitative information appears in the array of numbers below the illustration. The qualitative information is found in the *agreed significance* of the numbers.

In this chapter, we focus on *coding*: the representation and compression of the quantitative information.

4.1 Coding

A *code* (or, more precisely, a *binary code*) is a process in which numbers or other symbols are replaced (represented) by sequences of 0s and 1s called *code words*. The coding almost invariably serves purposes of storage or transmission of the original data. The inverse process, the reconstruction of these data, is called *decoding* or *deciphering*. In this process, one needs to know both the specific precepts according to which a data item is translated into a single code word (the 'code book') and the organization of these data. In the case of a digital image, for example, a knowledge of the number of rows and columns is required, as is the order of appearance of the image pixels.

EXAMPLE 4.2

When using Morse code, as is still done in telegraphy, the letters of the alphabet are replaced by sequences of dots and dashes. If, further, these are replaced by bits, for instance dots by 0 and dashes by 1, one has a binary coding of the alphabet. In this code, the code words for I, M, O, and L are 00, 11, 111, and 0100, respectively.

In practical applications, an extra symbol occurs in order to separate the individual code 'words' (that is, each letter) – effected as a pause terminating the transmission of a single letter. Morse code is, thus, not a binary code.

Also, in the so-called pulse code modulation (PCM) employed for storage of images, sound or data on digital tapes or compact discs, the binary symbols occur in groups. The spacing between groups is thus a separate symbol, and even if PCM is often referred to as binary coding, this is not the case, at least not in the strict sense of the word.

In what follows, the fundamental coding principles will be illustrated by means of proper binary codes. The above examples tell us that if the code words are to be separated from each other in an unambiguous way, the code must possess a specialized structure.

EXAMPLE 4.3
Table 4.1 lists three codings of the letters from I to P. Clearly, code 1 is not suitable for transmission of more than one 'code word', that is, one letter. For example, the sequence 1000110 may be decoded in many (how many?) ways – one is KIIJK. The sequence is not going to occur in code 2. In this code, however, any sequence with a number of bits equal to a multiple of 3 is allowed, and any such sequence – for example, 010110100 – will be unambiguously decipherable. Code 3 has the same property. From its code book, decode the sequence 1100011110010 for yourself.

The single feature ensuring unambiguous decoding is called the *prefix condition*: no word in the code book should appear as the first bits (a prefix) in any other word. When a code word has been received, there is no possibility that it is part of a longer one – the next bit must be the beginning of a new word. This property must be satisfied for all the codes to appear in what follows.

A code meeting the prefix condition may be viewed as a 'questioning strategy' in the following sense. If a single symbol among N symbols is to be encoded, this may be accomplished by asking yes/no questions in succession. If the answers are coded as 1 and 0, respectively, the symbol in question will always be uniquely identified, after which no further questions are asked. The resulting code book for the N symbols will thus satisfy the prefix condition.

EXAMPLE 4.4
Which of the 'codes' in Table 4.2 are, in fact, codes – bearing in mind the prefix condition?

Table 4.1 Codes for eight different symbols.

Letter	*Code 1*	*Code 2*	*Code 3*
I	0	000	01
J	1	001	00
K	10	010	10
L	11	011	1101
M	100	100	1100
N	101	101	11100
O	110	110	11101
P	111	111	11110

Table 4.2 Codes or not?

Symbol no.	*Code 1*	*Code 2*	*Code 3*	*Code 4*
1	00	001	1011	101
2	10	01	010	111
3	11	11	1001	001
4	010	011	111	010
5	0110	101	00	1100
6	0111	1001	0	0011

EXAMPLE 4.5

In a certain digital image, only five grey levels occur, called white, light grey, grey, dark grey, and black, respectively, for which we must choose code words. We now adopt the following questioning strategy:

Is the grey level white, light grey, or grey?

If yes:

Is the grey level white or light grey?

If yes:

Is the grey level white?

If no:

Is the grey level dark grey?

This strategy is converted into a code book as follows:

Grey level	white	light grey	grey	dark grey	black
Code	111	110	10	01	00

If the five levels occur equally frequently in the image, one will use, on average, $(3 + 3 + 2 + 2 + 2)/5$ bits $= 2.4$ bits per level. It seems plausible – and may be proved – that the coding cannot be realized with fewer bits. The chosen code is thus *optimal*, that is, it minimizes the number of bits. If some of the levels, however, appear more frequently than others, the code is not necessarily optimal. Such levels should be assigned short code words, as the average number of bits will be reduced accordingly.

4.2 Optimal coding

When considering reduction of transmission time and capacity, a fundamental question arises: What is the minimum number of bits needed to code a given amount of data?

Accordingly, we set out to find a lower limit for the number of bits necessary for the coding of N levels (or different symbols). The relevant quantity is the *average number of bits per level*. The situation occurring most commonly in practice is that of a large amount of data and code words of unequal length, since the various levels are known to occur with different frequencies. Even if these frequencies are alike, the code words will be equally long only if the number of levels is a power of 2. Only in this case will the average number of bits per level be equal to the actual word length.

Generally, the *information content* $\mathcal{J}$ *per level* is defined by

$$\mathcal{J} = \text{minimum average word length in code}$$

The quantity $\mathcal{J}$ depends only upon the number N of levels and their relative frequency. Thus, it denotes the minimum number of yes/no questions needed on average to specify any one of the N levels completely. A code of average word length $\mathcal{J}$ is called *optimal*, as mentioned.

For transmission of M data items, one needs $\mathcal{J}M$ bits; if, for instance, a $K \times L$ digital image is given, the number is $\mathcal{J}KL$ (cf. Example 1.4).

The first step is to demonstrate that, for $N = 2^n$ equally frequent levels, one has

$$\mathcal{J}(N) = \mathcal{J}(2^n) = n \tag{4.1}$$

where the dependence of $\mathcal{J}$ upon the number of levels N has been made explicit.

A possible strategy for specifying some level could start with asking: *Is the level to be found in the first half?* The answer reduces the subsequent number of levels to 2^{n-1}, and by posing the same question repeatedly the level is pinpointed after exactly n questions. If the N levels are represented by the numbers from 0 to $2^n - 1$, the resulting code book will consist of the binary representations of the individual numbers – where numbers having fewer than n binary digits are preceded by zeros. This code is called the *natural binary code* for the N levels (cf. code 2 in Example 4.3).

It is clear from the symmetry of the problem (that all the levels possess the same properties) that this code is optimal. Since the level is reached only after n questions, the number of bits is *always* n, and the information content becomes, as stated,

$$\mathcal{J}(N) = n = \log_2 N$$

where $\log_2$ denotes the logarithm of base 2.

If N is not a power of 2, the line of reasoning is as follows. First of all it is noted that the information content $\mathcal{J}(N)$ must increase with the number of levels, that is,

$$\mathcal{J}(N_1) \leqslant \mathcal{J}(N_2) \tag{4.2}$$

if $N_1 < N_2$. It is, of course, always possible to pin down one level among the smaller number N_1 if an optimal question strategy is available for the larger number N_2. But superfluous questions will inevitably be asked, and the strategy will not be optimal – hence the inequality (4.2).

Next, each of the N levels is subdivided into N new levels, a total of N^2 levels. These N^2 levels are coded by means of a strategy where the 'main' level is determined first, and the artificial sublevel thereafter, both steps using the same optimal questioning strategy. In this process, $\mathcal{J}(N) + \mathcal{J}(N)$ questions are asked. As this overall strategy for the N^2 levels is not necessarily optimal, we have

$$\mathcal{J}(N^2) \leqslant 2\mathcal{J}(N)$$

On subdividing M times instead of 2, we similarly see that

$$\mathcal{J}(N^M) \leqslant M\mathcal{J}(N) \tag{4.3}$$

For a given M we choose the power of 2 nearest to, but below, N^M:

$$2^{p(M)} \leqslant N^M < 2^{p(M)+1} \tag{4.4}$$

or

$$p(M) \leqslant M \log_2 N < p(M) + 1$$

that is, $p(M) = \text{int}(M \log_2 N)$, again denoting 'integral part' by int. From this inequality, one obtains, on dividing by M,

$$\frac{p(M)}{M} \leqslant \log_2 N < \frac{p(M)}{M} + \frac{1}{M}$$

This implies

$$\frac{p(M)}{M} \rightarrow \log_2 N \qquad \text{as } M \rightarrow \infty \tag{4.5}$$

But from inequalities (4.2) and (4.3), we also have $\mathcal{J}(N^M) \geqslant p(M)$ and, as this relation holds for all M, inequality (4.3) and expression (4.5) yield the important result

$$\mathcal{J}(N) \geqslant \log_2 N \tag{4.6}$$

supplementing equation (4.1) in those cases where N is not a power of 2:

The average number of bits when coding N equally frequent levels is at least $\log_2 N$.

On the other hand, if N is placed between two consecutive powers of 2 (that is, $p(1) = \text{int}(\log_2 N)$ and $p(1) + 1$, with the above notation), then

$$\mathcal{J}(N) \leqslant \text{int}(\log_2 N) + 1 \tag{4.7}$$

setting an upper limit on the number of bits required.

The results may be extended to the case where the levels do not occur equally frequently. Where large amounts of data are to be coded, the relevant concept is the *probability*, that is, the relative frequency of occurrence. The probabilities for the N levels are denoted $p_1, p_2, \ldots, p_N$ and referred to collectively as the *probability vector* $p = (p_1, \ldots, p_N)$. The average number of bits per level is, then, a function $\mathcal{J} = \mathcal{J}(N, p) = \Sigma\, p_n a_n$. Here, a_n stands for the number of bits (word length) for the nth level, coded optimally.

As we demonstrated above, a uniform distribution $p = (1/N, 1/N, \ldots, 1/N)$ leads to

$$\mathcal{J}(N, p) = \mathcal{J}(N) \geqslant \log_2 N$$

If, on the other hand, the ps are different (rational) numbers, we choose a large number K – the common denominator for the ps – so that all the values $K_n = Kp_n$ are integers; then $K = \Sigma\, K_n$, and these K levels may be regarded as equiprobable. They are coded by finding, first, the 'proper' level $n \in [1, N]$ by means of an optimal code and, next, the 'artificial' one among the K_n identical levels. The first step consumes, on the average, $\mathcal{J}(N, p)$ bits, the quantity sought, while the second step needs $\Sigma\, p_n \mathcal{J}(K_n)$. The total strategy is, however, not necessarily optimal; consequently

$$\mathcal{J}(K) \leqslant \mathcal{J}(N, p) + \sum_n p_n \mathcal{J}(K_n)$$

On using inequality (4.7) in the weaker form $\mathcal{J}(N) \leqslant \log_2 N + 1$ for the term $\mathcal{J}(K_n)$ we next obtain

$$\mathcal{J}(N, p) \geqslant \mathcal{J}(K) - \sum_n p_n \mathcal{J}(K_n) \geqslant \log_2 K - \sum_n p_n (\log_2 K_n + 1)$$

that is,

$$\mathcal{J}(N, p) \geqslant -\sum_n p_n \log_2 p_n - 1$$

As can be shown, this inequality may be strengthened to

$$\mathcal{J}(N, p) \geqslant -\sum_n p_n \log_2 p_n \tag{4.8}$$

The quantity on the right-hand side,

$$\mathcal{H}(N, p) = -\sum_{n=1}^{N} p_n \log_2 p_n \tag{4.9}$$

is called the *entropy* for the distribution p. Inequality (4.8) thus says:

> The minimal average number of bits needed to code N levels is greater than or equal to the entropy of the level probability distribution.

This main result is a generalization of the earlier one (p. 113). The example of the uniform distribution $p = (2^{-N}, 2^{-N}, \ldots, 2^{-N})$ shows that the inequality may not be strengthened further.

EXAMPLE 4.6
Show that, of all probability distributions on $[1, N]$, the uniform distribution $p = (1/N, \ldots, 1/N)$ is of maximum entropy.

EXAMPLE 4.7
Find the entropy for the distribution

(a) $(0.9, 0.1)$ (b) $(1/3, 1/3, 1/6, 1/6)$
(c) $(1/3, 1/3, 1/3)$ (d) $(1/2, 1/4, 1/8, 1/8)$

Write codes for each distribution and check that the average number of bits is larger than the entropy.

EXAMPLE 4.8
If p and q are probability vectors $p = (p_1, p_2, \ldots, p_M)$, and $q = (q_1, q_2, \ldots, q_N)$, their *product distribution*, pq, is that having probability vector

$$pq = (p_1 q_1, p_1 q_2, \ldots, p_M q_{N-1}, p_M q_N)$$

Show that $\mathcal{H}(MN, pq) = \mathcal{H}(M, p) + \mathcal{H}(N, q)$.

EXAMPLE 4.9
Demonstrate the fundamental result (4.8) by repeating the argument on pp. 113–14: the K artificial levels are subdivided successively into K^M levels, changing the N 'real' levels into N^M levels having probability distribution p^M (notation as in Example 4.7). First utilize $\mathcal{J}(p^M) \geqslant \mathcal{H}(p^M) - 1$, then Example 4.8, and, finally, the inequality $M\mathcal{J}(p) \geqslant \mathcal{J}(p^M)$.

In summary, the considerations in this section demonstrate the existence of a lower limit to the number of bits required to construct an unambiguously decipherable code (a code with no separation symbols) for a series of levels – for example, the grey levels in a digital image. This lower limit is given by the entropy for the statistical distribution of levels:

$$\mathcal{J}(N, p) \geqslant \mathcal{H}(N, p) = -\sum_{n=1}^{N} p_n \log_2 p_n$$

where N is the number of levels, p the probability vector for their distribution, $\mathcal{J}$ the minimal (average) number of bits per level, and $\mathcal{H}$ the entropy of the distribution.

Experience shows that the substitution of the information content $\mathcal{J}$ by its lower limit $\mathcal{H}$ is an excellent approximation, one that will be employed in what follows.

When constructing an optimal code for a given probability distribution, the so-called *Huffman coding* provides an elegant algorithm. The principle is illustrated in the next example.

EXAMPLE 4.10 (Huffman coding)
Construct an optimal code for the probability distribution

$$p = (0.15, 0.35, 0.07, 0.08, 0.20, 0.15)$$

Procedure. The smallest two probabilities, 0.07 and 0.08, are added, yielding a new distribution

$$p' = (0.15, 0.35, 0.20, 0.15, 0.15)$$

Iteration of this process leads to the following scheme:

```
0.15        0.15        0.15 --0--> 0.35        0.35 --0--> 1.00
0.35        0.35        0.35   /    0.35 --0--> 0.65 --1/
0.07 --0\   0.20        0.20 -1     0.30 --1/
0.20     >  0.15 --0--> 0.30
0.08 --1/   0.15 --1/
0.15
```

Where the minimal probability is represented three or more times, one is free to choose among them.

The principle for the coding is as follows. Going backwards, from right to left, one asks yes/no questions. At each question, a probability is split into two, and a 0 or a 1 (optionally) is written in the code book, corresponding to those probabilities involved in the splitting. With the choice made above, the result is as shown in Table 4.3.

It may be proved that the Huffman code is optimal. It is seen from Table 4.3 that the information content $\mathcal{J}(p)[=\mathcal{J}(6, p)]$ in the probability distribution p is

$$\mathcal{J}(p) = 0.15 \cdot 2 + 0.35 \cdot 2 + 0.07 \cdot 4 + 0.20 \cdot 2 + 0.08 \cdot 4 + 0.15 \cdot 3 = 2.45$$

In comparison, the distribution entropy is $\mathcal{H}(p) = -\Sigma\, p_n \log_2 p_n = 2.38$.

Table 4.3 Huffman code.

Probability	0.15	0.35	0.07	0.20	0.08	0.15
Code	00	10	1100	01	1101	111
Word length	2	2	4	2	4	3

EXAMPLE 4.11

Construct a Huffman code for the probability distribution p given by

$$p = (0.40, 0.32, 0.16, 0.08, 0.04)$$

and find its information content as well as its entropy.

EXAMPLE 4.12

A 1000×1000 digital image, containing five distinct levels, is to be transmitted in coded form, pixel by pixel. What is the minimal number of bits required, if

(a) no information is available concerning the level distribution,
(b) the levels are known to be distributed as follows?

Level no.	1	2	3	4	5
Number	200 000	400 000	100 000	100 000	200 000

Check, finally, that the number of bits derived is not in conflict with the lower limit set by the entropy of the level distribution.

Appendix B gives a BASIC subroutine calculating both entropy and information content for a given probability distribution by means of the Huffman coding principle.

4.3 Transmission of binary data

The above considerations show that, if a digital image is to be transmitted in coded form, a minimum number of bits is required, and the number is given by the entropy of the level distribution. This limitation is not going to cause problems if only the *transmission rate* (given in bits per second, also called *baud rate*) can be increased sufficiently. However, even in this case, limitations of principle arise, not to mention the practical limitations, as was discovered in the early days of telegraphy. The price to be paid for a higher transmission rate is, namely, an increased consumption of energy and a larger 'bandwidth' – notions to be investigated in this section.

When transmitting binary data, rectangular signals are employed, as mentioned earlier. Accordingly, we shall consider the passage of a step signal $s(t)$ through a transmission system – a rectangular signal being the difference between two step signals. For simplicity, it is assumed that the impulse response for the transmission system is a rectangular signal as well.

Here, the essential factor is not the rectangular shape as such; the important quantity is the *duration* τ of the response:

$$h(t) = \frac{A}{\tau} r_\tau(t)$$

The two signals in question are depicted in Figure 4.1.

The output of the transmission system is

$$g(t) = h \star f(t) = \int h(t')f(t-t')\,\mathrm{d}t'$$

$$= \begin{cases} 0 & t < 0 \\ At/\tau & 0 \leqslant t \leqslant \tau \\ A & t > \tau \end{cases}$$

as shown in Figure 4.2. The effect of the system is an amplification of the signal level 1 from the input to level A; this level is, however, not attained until time τ. This situation entails a problem. If the receiving part is to decide whether the signal value at the sender has increased from zero, this will only be possible after a certain interval of time has elapsed. The reason is that the receiver will not be able to distinguish the first (small) positive values received from zero. In practice, noise is bound to be present as well, so the receiver decision 'positive signal received' can, in reality, only be made when the signal exceeds a certain value, the *detection limit* d. Consequently, a certain *detection time* t_d will elapse:

$$t_d = \frac{d}{A}\tau \tag{4.10}$$

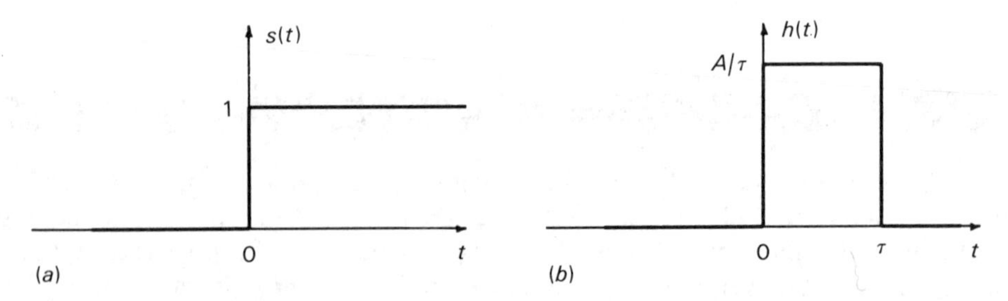

Figure 4.1 (a) Binary code (step) signal. (b) Impulse response.

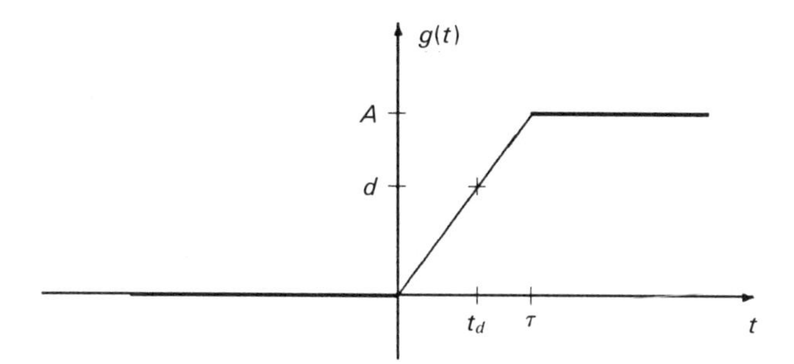

Figure 4.2 Transmitted binary code signal.

Clearly, this quantity imposes a lower limit upon the time interval between successive jumps in signal value at the sender, if no confusion is to arise at the receiving end.

Instead of using the quantity τ as a measure of the characteristic time for the transmission system, the system *bandwidth* Ω is often specified. As is seen from Example 3.13 and equation (3.47), the transmission system function obeys the following rule:

$$|H(e^{i\omega})| = \frac{2\sin(\tau\omega/2)}{\omega}$$

The frequencies for the harmonic signals which are not damped considerably when passing the system are thus concentrated in an ω-interval of approximate size

$$\Omega \simeq \frac{2\pi}{\tau}$$

This result may be shown to hold for all pairs of Fourier transforms in the sense that, loosely speaking, the characteristic 'life span' for a (transient) signal and its bandwidth are inversely proportional (cf. p. 83).

Since the number of bits per second must be in the neighbourhood of, or below, the quantity $1/t_d$, the restrictions upon the transmission rate of a system may be summarized in a relation of the form

$$\text{bit rate} \leqslant \left(\frac{A}{d}\right)\frac{1}{\tau} = \left(\frac{A}{d}\right)\frac{\Omega}{2\pi} \tag{4.11}$$

in which the amplification (A) and the detection limit (d) for the transmission system appear together with its response time (τ) or bandwidth (Ω).

If the relation (4.11) is compared with the above limit regarding the number of bits in a coding of N symbols or levels, occurring with frequencies $p = (p_1, p_2, \ldots, p_N)$, we find a minimum

$$\text{transmission time per level} \geqslant \left(\frac{d}{A}\right)\mathscr{H}(p)\tau = \left(\frac{d}{A}\right)\mathscr{H}(p)\frac{2\pi}{\Omega} \tag{4.12}$$

where $\mathscr{H}(p)$ is the entropy of the distribution, given by equation (4.9).

EXAMPLE 4.13

A TV-image with 600×500 pixels and 32 grey levels is to be transmitted in real time, that is, within 0.02 s. The detection ratio d/A is 0.1. Finally, of the 32 levels, 20 occur with probability 2 per cent, and the remaining 12 occur with probability 5 per cent.

Find the minimum number of bits needed to code this image, and the minimal bandwidth for the transmission system.

4.4 Quantization

Quantization was mentioned in Chapter 1 as part of the digitization process. When digitizing an image, it was divided into appropriately sized fields, and each field was assigned a value corresponding to the amount of light represented. The physical measure of this amount of light was the number of photons detected. The digital image was, then, a signal $b = b[m, n]$ giving the number of detections, or level, in pixel $[m, n]$; the digital position limits are $m \in [0, M-1]$ and $n \in [0, N-1]$.

In the quantization, the levels are binned, so that the total number of levels is reduced. To control this procedure, the *histogram* of the level distribution is an extremely powerful tool. In the histogram, we read off the number of times, $\#(k)$, that level k occurs in the image, and the probability of the level is thus

$$p_k = \#(k)/MN$$

since $MN = \Sigma_{k=1}^{K} \#(k)$ is the total number of pixels.

The histogram depends markedly upon the ratio of the number of levels and the number of pixels, K/MN. Just as was the case for the variation of the grain density with pixel size (Figure 1.4), three possibilities arise for the histogram, as indicated in Figure 4.3. In case (a) (where only a portion is shown), so many levels

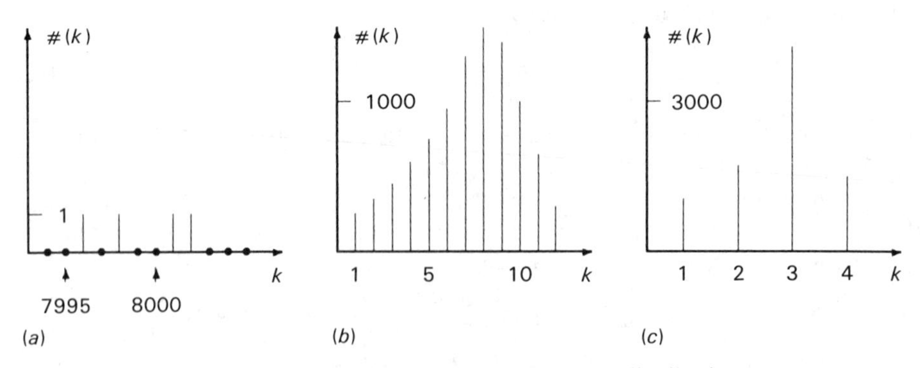

Figure 4.3 Three histograms for the same distribution.

exist that only a small fraction is represented and, if so, presumably only once. Histogram (b) is created from (a) by grouping of successive levels. The process has been carried one step further in histogram (c). Thus, the histograms are created by successive quantizations.

It seems plausible that histogram (b) is the most interesting one (even if (a), in principle, contains the complete information). The 'quality criterion' is, once again, that the histogram values do not change appreciably when going from one value to its neighbours (cf. Sections 2.2 and 2.3).

Normally, therefore, a certain freedom exists when choosing the degree of detail to be represented in a histogram, even if the template shown in Figure 4.3b is prescribed. Any choice, however, invariably means a compromise between, on one hand, the conservation of the original image information and, on the other hand, the possibility of coding this information with a conveniently small number of bits. We now attempt to quantify these conflicting options, as we compare a fine and a coarse quantization (cf. Figure 4.4).

Figure 4.4a shows a portion of an image histogram, constructed as a probability distribution, corresponding to a definite quantization with K levels. In the extra horizontal l-axis, some of the L new histogram points, $l = 1, 2, \ldots, L$, have been marked; the broken vertical lines mark the corresponding quantization boundaries. The new histogram values q_l, $l \in [1, L]$, are constructed from the old ones, $p_k, k \in [1, K]$, as follows:

$$q_l = \sum_{k=k_l^-}^{k_l^+} p_k$$

that is, by summing the probabilities p_k within the quantization boundaries in group no. l, denoted k_l^- and k_l^+, respectively. A portion of this histogram is shown in Figure 4.4b. For instance, this figure shows that $k_{14}^- = 89$, $k_{14}^+ = 91$, $k_{15}^- = 92$, etc.

In a certain sense, it is possible to reconstruct the histogram for the finer quantization, as the 'old' probabilities are simply replaced by their average values

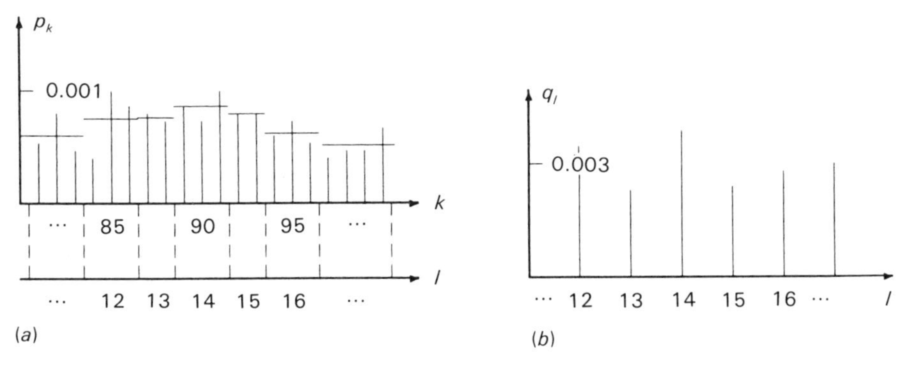

Figure 4.4 Successive quantization.

over the groupings. This histogram is suggested by the horizontal lines in Figure 4.4a. The procedure is quite similar to the reconstruction of an analog signal from a digital version thereof (cf. Figure 2.17). Note that if the new quantization were carried out, starting from the reconstructed one, the resulting histogram would once again be Figure 4.4b.

If the lth new quantization level consists of $n_l = k_l^+ - k_l^- + 1$ old levels (where $\Sigma_l\, n_l = K$), then the K reconstructed probabilities $\bar{p}_k$ are given by

$$\bar{p}_1 = \bar{p}_2 = \cdots = \bar{p}_{n_1} = \frac{1}{n_1} \sum_{k=1}^{n_1} p_k = \frac{q_1}{n_1}$$

and similarly for the remaining $L-1$ groups.

We next calculate the entropy in the original distribution by approximating it with its reconstruction $\bar{p}_k$. We thus approximate $\mathcal{H}(N, p)$ with $\mathcal{H}(N, \bar{p})$:

$$\mathcal{H}(N, \bar{p}) = -\sum_{k=1}^{K} \bar{p}_k \log_2 \bar{p}_k = -\sum_{l=1}^{L} n_l (q_l/n_l) \log_2 (q_l/n_l)$$

since the first sum consists of n_1 equals terms $(q_1/n_1)\log_2(q_1/n_1)$, then n_2 equal terms, etc. Consequently

$$\mathcal{H}(N, p) \approx \mathcal{H}(N, \bar{p}) = -\sum_{l=1}^{L} q_l \log_2 q_l + \sum_{l=1}^{L} q_l \log_2 n_l$$

In this sum, the first term is the entropy of the new distribution:

$$\mathcal{H}(L, q) = \sum_{l=1}^{L} q_l \log_2 q_l \tag{4.13}$$

while the second term is the mean value of $\log_2 n_l$ taken over the new distribution:

$$E(\log_2 n_l) = \sum_{l=1}^{L} q_l \log_2 n_l \tag{4.14}$$

The result of these considerations is that the coarser quantization implies an information loss

$$\mathcal{J}(K, p) - \mathcal{J}(L, q) = \mathcal{H}(K, p) - \mathcal{H}(L, q) = E(\log_2 n_l) \tag{4.15}$$

with a corresponding decrease in number of bits needed for coding.

EXAMPLE 4.14

An image has been quantized with six levels, the frequencies of which are listed below.

Level	1	2	3	4	5	6
Probability	0.05	0.10	0.15	0.20	0.25	0.25

It is to be quantized with only three levels instead. What are the various possible information losses? Find the quantization corresponding to (a) maximum loss, (b) minimum loss.

On the other hand, a coarser quantization will introduce an error in the sense that the distance is increased between the original image – whether itself quantized or not – and the image corresponding to the new, coarser quantization. What is the error introduced in this manner? This question will be addressed in the remainder of this section.

The simplest quantization process is that of replacing certain consecutive levels by one new level. As above, we assume that these levels are equiprobable, and, for simplicity, we first estimate the error arising from binning the integers from 0 to $N-1$. Here, $p_0 = p_1 = \cdots = p_{N-1} = 1/N$, and the mean value of the distribution is

$$E(n) = \sum_{n=0}^{N-1} \frac{1}{N} n = \frac{1}{N} \sum_{n=0}^{N-1} n = \frac{N-1}{2}$$

and as

$$E(n^2) = \sum_{n=0}^{N-1} \frac{1}{N} n^2 = \tfrac{1}{6}(N-1)(2N-1)$$

(cf. Example 3.29), the variance becomes

$$\begin{aligned} V(n) = E(n^2) - E(n)^2 &= \tfrac{1}{6}(N-1)(2N-1) - \tfrac{1}{4}(N-1)^2 \\ &= \tfrac{1}{12}(N^2-1) \end{aligned} \tag{4.16}$$

In replacing the N levels by their mean value $(N-1)/2$, we obtain a quantization error which, in the mean, equals zero. The distance notion introduced in Section 2.5 for signals and images did, however, involve a sum of squares of the individual deviations. This deviation is, in the mean, given by equation (4.16).

If, instead, we consider the n_l levels $k \in [k_l, k_l + n_l - 1]$, the mean value becomes

$$E(k) = k_l + \frac{n_l - 1}{2}$$

whereas the variance does not change as a result of the shift; the value $(n_l^2 - 1)/12$ is retained. The image distance between the original image b and the quantized image $\bar{b}$ (where the original image, too, will in general be quantized) is thus given by

$$|b - \bar{b}|^2 = \sum_{m=0}^{M-1} \sum_{n=0}^{N-1} (b[m,n] - \bar{b}[m,n])^2 = \sum_{l=1}^{L} MNq_l \frac{n_l^2 - 1}{12}$$

where use has been made of the fact that, of the MN image pixels, MNq_l are quantized at the same level. Thus, the deviation between the two images is specified by the distance measure (cf. p. 55)

$$\frac{1}{MN}|b-\bar{b}|^2 = \sum_{l=0}^{L-1} q_l \frac{n_l^2-1}{12} = E(n_l^2-1)/12 \qquad (4.17)$$

In summary:

Quantization of an original image with levels $[1, K]$ into a new one of L $(< K)$ levels, in which every new level consists of $n_l, l \in [1, L]$, original ones, the coding advantage is

$$E(\log_2 n_l)$$

and the image distance loss

$$\sqrt{E(n_l^2-1)/12}$$

per pixel, where the mean values should be taken over the new level distribution.

EXAMPLE 4.15

Find the losses in image distance for the two extremes among the new quantizations of the distribution in Example 4.14.

In many practical cases, the groups contain the same number of levels, $n_1 = n_2 = \cdots = n_L = K/L$, so that the mean value symbol above may be omitted. Also, $n_l = K/L$ will often be a 'large' number, and we may disregard the '-1' in the expression for the image error. If so, Table 4.4 applies. These results are easily generalized to the case where the original K levels are not, necessarily, consecutive integers.

Reminding ourselves briefly of the rigorous view of a digital image as a set of stochastic variables b (one for each pixel), every image has an 'inherent' error, comparable to a quantization error. If, for instance, the digital image consists of photon numbers behaving according to a Poisson distribution, and if the ideal photon number is a given pixel (the detection number per unit time in the limit of

Table 4.4 Quantization: for and against.

Coding advantage per pixel	*Image error per pixel*
$\log_2(K/L)$	$(K/L)/\sqrt{12}$

an infinitely long exposure equals the mean value for the Poisson distribution corresponding to the pixel in question) is b, then the number observed will fluctuate between approximately $b - \sqrt{b}$ and $b + \sqrt{b}$. Normally, therefore, there will be no reason to quantize with more levels than, approximately, $\sqrt{b}$, around level b.

This may also be stated by saying that the 'natural' image error is given by the square root of

$$\frac{1}{MN} \sum_m \sum_n V(\mathrm{b}[m, n]) \simeq \frac{1}{MN} \sum_m \sum_n b[m, n] = E(p) \tag{4.18}$$

that is, the mean value of all photon numbers – or, alternatively, the histogram mean value.

We note in passing that it may prove statistically convenient to employ a level-*dependent* quantization. If, for example, the number of levels in a group near the level b is chosen to be proportional to $\sqrt{b}$, one obtains, approximately, the same combined image error in all the pixels.

The quantization error will normally be quite small in comparison with the natural error deriving from the Poisson distribution in the individual pixels. Typically, the quantization accuracy will thus be dictated by the storage or transmission capacity available.

EXAMPLE 4.16

In a certain digital image, photon counts from 1 to 9 are encountered. The frequencies are

$$p_n = n(10 - n)/165$$

Find the distribution entropy and the information content as well as the natural image error per pixel.

This image is quantized with the groupings

(a) 1 23 456 78 9
(b) 123 456 789

Answer the same questions in both cases, replacing the 'natural image error' by the 'quantization error'. The latter may conveniently be expressed as multiples of the former.

4.5 Segmentation

When an image is *segmented*, it is divided into several areas, each of which is considered to be a new separate image. Usually, the resulting image segments are assumed to be *connected*, that is, any two pixels within a segment must be 'in contact' via neighbouring pixels. Figure 4.5 is an image with four segments. The segment in the lower left-hand corner consists of only one pixel and its four neighbouring pixels. The segment in the middle of the image has been equipped with a chain of adjoining pixels, connecting pixels A and B. The grouping on the right is non-connected and should be regarded as two separate segments.

Figure 4.6 is a more realistic example. Figure 4.6a is unequivocally perceived as consisting of an object and a background. This is confirmed by the histogram Figure 4.6b, also consisting of two components: a portion (right) with a reasonably smooth distribution of levels (corresponding to the object) and a concentrated distribution corresponding to the background (left). The segmentation of this image, isolating the object, is thus reflected in the partitioning of the histogram into two parts.

The usefulness of such a partition lies in its 'objectiveness' – it may, for instance, be carried out by computer. This is done according to the principle that the histogram levels are divided into groups with distinct statistical properties. Next, the image is segmented into the areas corresponding to the histogram groups.

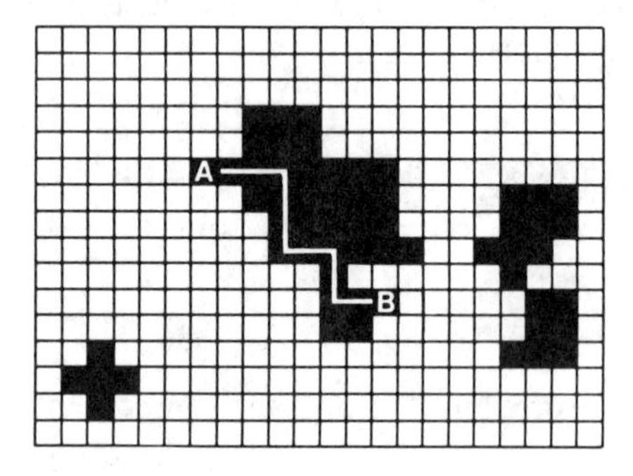

Figure 4.5 Image with four segments.

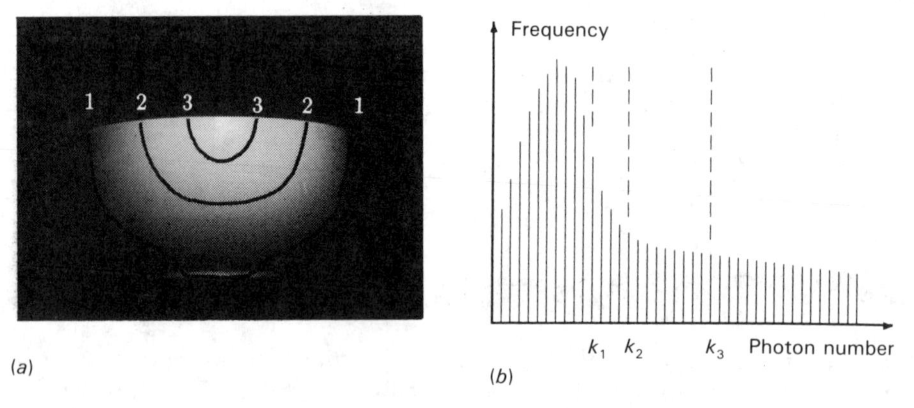

Figure 4.6 Image and its histogram (stylized).

Figure 4.6a shows the effect of three such segmentations, where the thresholds (k_1, k_2, and k_3) are marked in the histogram. The corresponding image segments are indicated by the three contour curves.

Apart from the purely practical aspects of image segmentation, large information reductions result from this procedure – especially, of course, if irrelevant image areas are simply discarded. The typical situation is as outlined below.

An image with levels $k \in [1, K]$ is split into two segments according to the level partition $k \in [1, L]$ and $k \in [L+1, K]$. The total image entropy is

$$\mathscr{H}(K, p) = -\sum_{k=1}^{L} p_k \log_2 p_k - \sum_{k=L+1}^{K} p_k \log_2 p_k$$

The two sums are *not* segment entropies, since the sum of the ps within the groups is not unity; thus the ps are not probability distributions. However,

$$\begin{aligned} P_k &= p_k/P^{(1)}, \qquad k \in [1, L] \\ P_k &= p_k/P^{(2)}, \qquad k \in [L+1, K] \end{aligned} \tag{4.19}$$

are probability distributions if $P^{(1)} = \Sigma_{k=1}^{L} p_k$ and $P^{(2)} = \Sigma_{k=L+1}^{K} p_k$. Note that $(P^{(1)}, P^{(2)})$ constitutes a probability distribution as well. Then

$$\begin{aligned} \mathscr{H}(K, p) &= -\sum_{k=1}^{L} P^{(1)} P_k \log_2(P^{(1)} P_k) - \sum_{k=L+1}^{K} P^{(2)} P_k \log_2(P^{(2)} P_k) \\ &= P^{(1)} \mathscr{H}_1 + P^{(2)} \mathscr{H}_2 + \mathscr{H}(2, P) \end{aligned} \tag{4.20}$$

Thus, the original image entropy has been written as a sum of three contributions, one from each of the segments ($\mathscr{H}_1$ and $\mathscr{H}_2$) besides the 'segmentation entropy' $\mathscr{H}(2, P)$. The former two terms,

$$\mathscr{H}_1 = -\sum_{k=1}^{L} P_k \log_2 P_k \qquad \text{and} \qquad \mathscr{H}_2 = -\sum_{k=L+1}^{K} P_k \log_2 P_k$$

are also called the *conditional entropies*. The segmentation entropy is

$$\mathscr{H}(2, P) = -P^{(1)} \log_2 P^{(1)} - P^{(2)} \log_2 P^{(2)}$$

If the original image consists of N pixels, the two segments consist of $N_1 = NP^{(1)}$ and $N_2 = NP^{(2)}$ pixels, respectively. To code the original image, a minimum number of bits given by

$$N\mathscr{H} = N_1 \mathscr{H}_1 + N_2 \mathscr{H}_2 + N\mathscr{H}(2, P)$$

is required. If, instead, the image is coded within segments, one needs only $N_1\mathscr{H}_1$ and $N_2\mathscr{H}_2$ bits, respectively. The reduction is thus to be found in the term $N\mathscr{H}(2, P)$. If, moreover, one segment (for example, no. 2) is superfluous, the corresponding term $N_2\mathscr{H}_2$ may be discarded.

EXAMPLE 4.17
A 1000×1000 digital image consists of 64 grey levels, 12 of which occur with probability 1/18 and the remaining 52 with a probability 1/156 each. The image is segmented according to this partition. With the previous notation, $P_1 = 12/18 = 2/3$ and $P_2 = 52/156 = 1/3$. The image entropy is

$$\mathcal{H}(p) = -12 \tfrac{1}{18} \log_2 \tfrac{1}{18} - 52 \tfrac{1}{156} \log_2 \tfrac{1}{156} = 5.21$$

and the entropies of the two segments are $\mathcal{H}_1 = \log_2 12 = 3.585$ and $\mathcal{H}_2 = \log_2 52 = 5.70$. The segmentation entropy is

$$\mathcal{H}(P) = -\tfrac{2}{3} \log_2 \tfrac{2}{3} - \tfrac{1}{3} \log_2 \tfrac{1}{3} = 0.92$$

Thus, coding of the original image requires of the order of 5 million bits. Segment no. 1 requires only approximately 2.4 million bits, segment no. 2 about 1.9 million bits. If segment no. 2 is the one containing the interesting information, a capacity reduction of over 60 per cent is obtained.

In the coding of an image segment, a description of the segment boundary is needed as well. Figure 4.7 shows, once again, a two-component image (with object and background). This figure suggests one among several ways of characterizing the location of a segment in the original image. Here, the image is 'scanned' upwards from below, row by row ($n \in [0, N-1]$) and from left to right within each row $[m, n]$, $m \in [0, M-1]$. For each n, we specify the smallest value of m for which the pixel $[m, n]$ belongs to the segment; next, we give the number of adjoining pixels in the row $[m, n]$, $[m+1, n]$, ... located within the segment. As is shown, multiple 'hits' in the same row are possible.

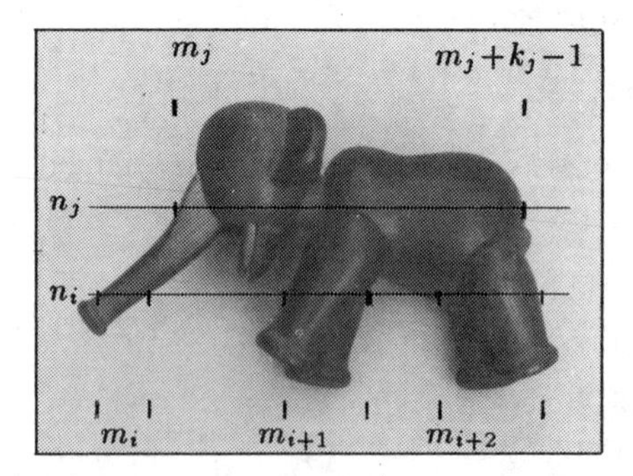

Figure 4.7 Image segment and boundary specification.

The results are collected into a table:

Row no.	n_1	n_2	...
Segment from column no.	m_1	m_2	...
Number of segment pixels in row	k_1	k_2	...

This table may, if one wishes, be regarded as a digital image and may be coded accordingly.

EXAMPLE 4.18
Many image segments, for example those which are *convex*, have the property that a row number n occurs only once, if at all. If so, the first line in the above table will be superfluous. If the vertical size of the segment is L ($\leqslant N$), the table may be regarded as a $2 \times L$ digital image having M levels. This 'image' may be coded using approximately $2N \log_2 M$ bits. If, for instance, $M = N = 1000$, this number is only about 20 000 bits.

Another possibility is to introduce a code for the segment contour. If the contour is regarded as belonging to the segment, it may, by force of the connectedness, be viewed as an unbroken chain of neighbouring pixels, as shown in Figure 4.8. The contour is completely specified by a sequence leading from one contour pixel to its neighbour. Only four possibilities exist (right, up, left, down); each one corresponds to one of the complex unit roots w_4^n of fourth order, that is 1, i, -1, or $-$i. The contour may then be coded as a one-dimensional digital (complex) signal, the components of which are of the form $A_n w_4^n$. This signal is obviously periodic.

Interestingly, a DFT of this 'contour signal' yields a representation (a so-called *Fourier descriptor*) which is independent of position, orientation, and size of the original segment.

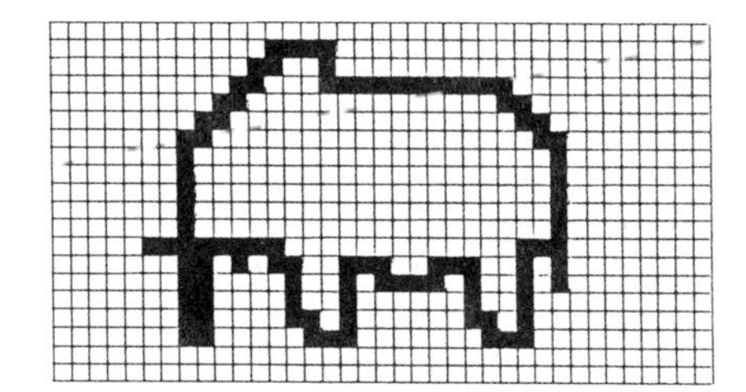

8i, −1, i, −1, i, −1, i, −10, 2i, −3, −i,
−1, −i, −1, −i, −1, −i, −1, −i, −1,
−6i, −2, 2, −5i, 1, 5i, 2, −i, i, 2, −i,
1, −3i, 1, −i, 2, 4i, 2, −i, 3, i, 2, −3i,
1, −i, 2, 5i, 2, −2i, 8i, ...

Figure 4.8 Image segment, contour, and contour code.

EXAMPLE 4.19
The segment in Figure 4.8 is rotated 90° anticlockwise and moved one pixel upwards. Find the code for the contour of the rotated segment.

EXAMPLE 4.20
An image segment consists of N pixels which have been coded as $f[n]$, $n \in [0, N-1]$, as described above. The Fourier descriptor is called F. Show that $F[0] = 0$.

4.5.1 Optimal segmentation

Finally, we wish to investigate how an image histogram may best be partitioned – that is, how to separate image segments from each other as clearly as possible. The discussion is restricted to a simple case where the image is assumed to consist of one or more segments within which the intensity is constant (Figure 4.9a). The ideal image thus consists of only two grey levels, the background level b and the object level c; however, due to statistical fluctuations, the histogram looks like that shown schematically in Figure 4.9b. This histogram must be a linear combination of two Poisson distributions:

$$p_n = P\mathrm{e}^{-b}\frac{b^n}{n!} + Q\mathrm{e}^{-c}\frac{c^n}{n!} \tag{4.21}$$

Here, P and Q denote the relative number of pixels (cf. equations (4.19)), so that $P + Q = 1$. We now place a partition d between b and c and attempt to choose this

(a)

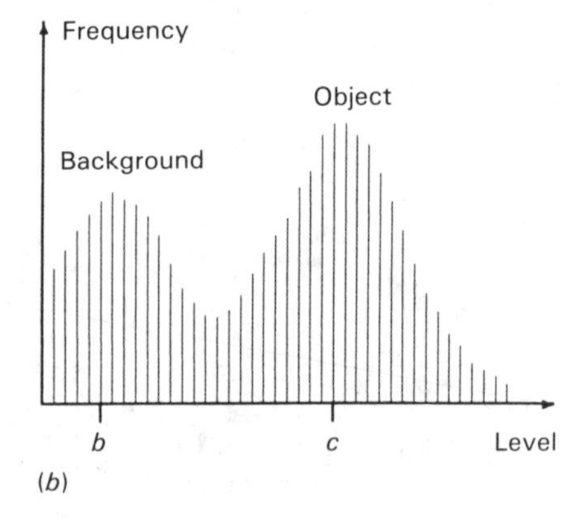

(b)

Figure 4.9 Image with two intensities.

threshold in such a way as to render the corresponding segmentation the best possible. No matter how d is chosen, a certain probability of misidentification of pixels will remain. The probability of classifying a background pixel as an object pixel is

$$\sum_{n=d}^{\infty} e^{-b} \frac{b^n}{n!}$$

and the probability of classifying an object pixel as a background pixel is

$$\sum_{n=0}^{d-1} e^{-c} \frac{c^n}{n!}$$

Thus, the total probability for misclassification is

$$p_f = Pe^{-b} \sum_{n=d}^{\infty} \frac{b^n}{n!} + Qe^{-c} \sum_{n=0}^{d-1} \frac{c^n}{n!}$$

We next investigate the sensitivity of this probability to the chosen threshold. Changing d to $d+1$ will result in a change in p_f equal to

$$\Delta p_f = -Pe^{-b} \frac{b^d}{d!} + Qe^{-c} \frac{c^d}{d!}$$

This quantity vanishes on choosing

$$d_0 = \frac{c - b + \ln(P/Q)}{\ln(c/b)} \tag{4.22}$$

so that the optimal threshold is given by $d = \mathrm{int}(d_0)$.

In practice, the relative number of levels will depend upon the threshold. However, it is often possible to choose a reasonable initial value d, after which the number of levels in the two groups may be evaluated. Next, the optimal value for d is found, corresponding to these numbers of levels, and the process is repeated. Normally, this iterative procedure converges very rapidly, leading to the final solution to the optimization problem.

5

Detection, Recognition, and Estimation

Viewed in the context of our main theme of 'information reduction', the three topics of the present chapter provide extreme cases. For instance, in a detection process, an entire image (or, more often, an image segment) is reduced to one bit....

The purpose of a *detection* is to answer a question of the type: Does the image contain something of interest? The 'something of interest' could be an object, often of an unknown nature, distinguishing itself from the background of the image – that is to say, the object is recognizable in spite of the image noise. Another common situation is the presence of special objects in a background of other – currently uninteresting – objects. Here, one is expected to solve a problem such as 'What is wrong with this picture?' In any case, the task is to decide whether an image content is 'normal' or whether something unexpected is present.

Thus, the detection process is in effect a statistical test; clearly, the image distance measure introduced in Section 2.5 will play a decisive role in this context. The concept of detection described in this chapter may be regarded as a generalization of the elementary photon detection (Chapter 1), where the 'object' is a (recorded) photon – here, the detection consists in the fact that the photon number in the pixel under consideration has been increased from n to $n + 1$.

In a *recognition* process, a somewhat higher degree of differentiation is called for, since one is further expected to *classify* a detected object, that is, to perform a comparison with certain reference objects. Here, the degree of similarity between the detected object and the reference objects is quantified and, if deemed satisfactory, the object is identified. Also in this situation, the procedure is a statistical test, the outcome of which may be either positive (with an unambiguous identification among the candidates at hand), negative (no certain identification), or, possibly, inconclusive (two or more possible identifications).

Finally, the concept of *estimation* relates to the evaluation of *image parameters*, that is, quantities considered to be relevant to the characterization of the objects in an image: dimension, shape, total intensity, and so on. The generic term *features* is often used to describe such quantities.

EXAMPLE 5.1

The images below illustrate the three processes.

If Figure 5.1a is compared with Figure 5.1b, it seems obvious that an object is *detected*, in the above sense, in Figure 5.1b. A comparison with the reference objects below allows a classification, *recognition*, of the object – although the task in practice surely will be less straightforward! Finally, the ruler above the object indicates an *estimation* of its extent.

A very rough evaluation of the information gained from the three processes might be as follows:

Process	Detection	Recognition	Estimation
Information content (bits)	1	$\log_2 4 = 2$	$\log_2 20 = 4.32$

In the estimation case, we have arbitrarily assumed that the 20 possible readings of the ruler are equiprobable.

(a) (b)

Figure 5.1 Detection, recognition, and estimation.

5.1 Pointwise detection

The simplest possible detection problem is to decide whether the intensity in a given pixel deviates from a certain 'expected' value. If this expected level is b, the observed level n must be Poisson distributed with parameter b:

$$p_n = e^{-b}\frac{b^n}{n!}$$

as shown in Section 1.5. As a measure of the quality of this detection, one would normally specify the number of standard deviations between the observed photon number and the expected one:

$$D = \frac{|n-b|}{\sqrt{b}} \tag{5.1}$$

In practice, the detection is based upon the prior choice of a threshold value D_0 – normally 1, 2, or 3. If the D observed is greater than (or equal to) D_0, this is called a detection; if D is smaller, the observation is not regarded as a manifestation of anything unusual. If D_0 is chosen as above, we speak of, respectively, a *one-sigma*, *two-sigma*, or *three-sigma* detection, with reference to the traditional symbol σ for the standard deviation. The thresholds are

$$\left.\begin{matrix} n_+ \\ n_- \end{matrix}\right\} = b \pm D_0\sqrt{b} \tag{5.2}$$

In order to ensure a systematic use of the detection notion, it is essential that both the pixel under consideration and the threshold be chosen *before* the observation, a normal practice in all statistical tests. As is well known, it is meaningless to calculate the probability of obtaining a 'six' with a die *after* the event – or, for that matter, the probability of reading exactly this paragraph at this very moment.

In image analysis, the tests to be performed will often be *one-sided*, that is, the decision to be made regards whether the photon number *exceeds* an expected value. If so, we shall only ask whether the observed photon number n falls above or below the upper detection limit $n_+ = b + D_0\sqrt{b}$.

EXAMPLE 5.2

A 1000×1000 digital image is the result of exposure with uniform illumination: in each pixel, a mean value of $b = 100$ photons is recorded. A 5σ detection *in a pre-chosen pixel* corresponds to at least $100 + 5 \cdot 10 = 150$ photons. The probability of this outcome is

$$e^{-100}\sum_{n=150}^{\infty}\frac{100^n}{n!} = 1.88 \times 10^{-6}$$

It should not, however, prove difficult to *pick* pixels with $n \geqslant 150$. Calculate the probability of finding at least one such pixel.

When testing a statistical hypothesis, two types of erroneous conclusion exist. First, the hypothesis may be rejected in spite of its being true (a 'type I error'); second, it may be accepted in spite of its being false (a 'type II error'). In signal and image processing, these two types of error are referred to as *non-detection* and *false alarm*, respectively.

In the one-pixel detection, the probability of false alarm, that is, a detection caused by statistical fluctuations, is given by

$$p_f = e^{-b} \sum_{n=n_+}^{\infty} \frac{b^n}{n!}$$

To reduce this probability, n_+ must be increased. This, however, increases the probability of non-detection, that is, the situation that a physically increased photon number – because of fluctuations – falls below the threshold.

Because of these conflicting requirements, an optimal value of D_0 will (in general) exist which minimizes the total probability of error. To determine this value, one needs to know the distribution of excess (object) photons $n - b$. Moreover, a prior knowledge of the occurrence of detections is necessary (the *a priori* probability).

If the object is expected to produce (in the mean) b' extra photons, the situation is identical to that discussed in Section 4.5.1, and the optimal threshold may be evaluated from equation (4.2.2), where Q should be regarded as the a priori probability of detection (whence $P = 1 - Q$). See also Figure 4.9b where, however, the 'object peak' should be reduced considerably to describe a realistic detection problem.

EXAMPLE 5.3

If the normal photon number is $b = 200$, the extra photon number $b' = 10$, and the a priori probability Q of detection is 50 per cent, then $n_+ = 204$. Find the optimal threshold in the following cases:

Normal photon number b	Extra photon number b'	Q (%)
2000	1000	40
10 000	100	40
10 000	1000	5

As will be evident from the foregoing, the Poisson distribution is rather awkward to use in practice, since it entails evaluation and summation of the quantity

$b^n/n!$ – the factorial is rather inconvenient, especially for large n. Another problem is the fact that an image described by the Poisson distribution may be subjected to neither arithmetic nor geometric manipulations (such as transformations) if the analytical knowledge of the statistical properties is to be retained. An exception is, of course, image *additions*, since the sum of two Poisson variables is Poisson itself.

If, however, the Poisson distribution is approximated by the normal distribution, as mentioned towards the end of Appendix A, this vital property is conserved. Moreover, the normal distribution is far more pleasant computationally; for many of the fundamental image manipulations, the resulting image statistics will be known explicitly. Finally, as regards the detection problem, the detection thresholds will often be established empirically, in which case the normal distribution frequently will be the natural choice. The approximation is given, for convenience, in Table 5.1.

EXAMPLE 5.4
What is the probability of false alarm when using a 3σ detection limit in a normal distribution? (See Appendix A.)

EXAMPLE 5.5
Compute $p_n = e^{-b}b^n/n!$ by means of the normal distribution in the following cases:

Case no.	1	2	3	4
b	10	100	1000	10 000
n	7	115	1000	10 090

and compare the result with Figure 1.25 (p. 29).

Most frequently, the actual photon numbers are not available, since they have been converted into other quantities. This may happen during readout and A/D conversion from a CCD chip and may be due to performed digital image

Table 5.1 Approximation of the Poisson distribution.

Poisson distribution	*Normal distribution*
$p_n = e^{-b}\dfrac{b^n}{n!}$	$p_n \approx \dfrac{1}{\sqrt{2\pi b}}e^{-(n-b)^2/2b}$

manipulations as well. In this case, too, the image statistics will often be described by the normal distribution, so the level x in the pixel under consideration has a probability density function

$$f(x) = \frac{1}{\sqrt{2\pi}\sigma} e^{-(x-b)^2/2\sigma^2}$$

where b once again represents the 'ideal' pixel value. The variance σ^2 is, however, no longer equal to b.

EXAMPLE 5.6
The expected level in a certain pixel is 10 000, and the standard deviation is 20. An observation of a suspected object is carried out five times, with the results 10 028, 10 018, 10 008, 10 021, and 10 019, respectively. Is this a case of 2σ detection?

EXAMPLE 5.7
In a certain pixel, where the expected level is b and the standard deviation equals σ, an extra intensity component appears, with an a priori probability Q; its mean is b' and its standard deviation is σ'. The level distribution is thus approximated by

$$f(x) = \frac{P}{\sqrt{2\pi}\sigma} e^{-(x-b)^2/2\sigma^2} + \frac{Q}{\sqrt{2\pi}\sigma'} e^{-(x-b-b')^2/2\sigma'^2} \tag{5.3}$$

Find the optimal detection threshold in the case $b = 1000$, $\sigma = 100$, $b' = 100$, $\sigma' = 10$, and $Q = 0.1$.

5.2 Area detection

The above methods for detection of an abnormal number of photons in a certain pixel may, with very little modification, bc applied to the more realistic case where the detection is to take place over a finite image area. However, certain differences of principle remain.

The most important property to be copied is the following. If two or more pixels in an image are grouped into a single one (for example, in order to reduce the information content or to eliminate noise), the value in the new, larger pixel will follow the same distribution as those in the individual ones, as the Poisson distribution and the normal distribution are both conserved by addition. It follows that the theory developed for one-pixel detection is equally applicable to detection of excess photons over an area consisting of more pixels, if the corresponding individual numbers are added.

If, on the other hand, the task is to detect one deviating pixel over a large image area of N pixels, the situation is somewhat changed. Assume, for simplicity, that the detection thresholds in each pixel have been placed so as to obtain the same probability, p, of avoiding false alarm in all N pixels. The probability of no false alarms at all across the entire area is then p^N – a number which, in general, will be much smaller than p. If, then, a probability p of avoiding false alarms over the entire area is to be achieved, the threshold for the individual pixels must be increased, that is, p should be replaced by

$$p' = p^{1/N} \tag{5.4}$$

EXAMPLE 5.8

In a 10×10 image area, the probability of false alarm is, in each pixel, 1 per cent. Calculate the probability of false alarm over this area, that is, the probability that at least one false alarm occurs somewhere. If the area probability is required to be 1 per cent, what should the one-pixel probability be?

The admittedly not very favourable detection circumstances outlined above are improved markedly if one is to detect objects as such, that is, image segments. This is caused by the fact that if some pixel flashes 'detection', its neighbouring pixels are no longer 'arbitrary' in the statistical sense, and the relevant detection limit is, once again, the one-pixel threshold. The same holds if two or more images of the same scene are available. Here a detection in some pixel renders that pixel atypical in the other images. This is exactly what happens when we try to locate a faint star in the sky, since the short 'exposure time' of the eye (approximately 0.05 s) facilitates a comparison between many images of the sky area containing the candidate object.

For the purpose of area detection, the statistical properties characterizing each pixel in the area must be known. Quite often, we shall make the fundamental assumption that *the statistical properties of the image are constant across the area*, that is, that all pixels are described by the same statistical distribution (cf. Section 5.4).

A necessary condition is, thus, that the mean value is constant across the area – the area must be uniformly illuminated. This requirement seems at first sight to exclude all realistic situations, such as that suggested in Figure 5.1, where the detection of an object takes place among other objects – the scene contains intensity variations. This difficulty is remedied by *background subtraction*; if we, for instance, subtract the digital image in Figure 5.1a from that in Figure 5.1b, the result is a constant background, of mean zero everywhere. Here, the detection amounts to a test of the hypothesis 'image = 0' against the alternative 'image > 0'.

If the object image is called $b_{\text{obj}}[m, n]$ and the background image $b_{\text{bgr}}[m, n]$, we assume these quantities to be normally distributed with mean values $\bar{b}_{\text{obj}}[m, n]$ and $\bar{b}_{\text{bgr}}[m, n]$ (the 'ideal images' corresponding to infinitely large intensities).

Correspondingly, the variances may be regarded as images $\sigma^2_{obj}[m,n]$ and $\sigma^2_{bgr}[m,n]$, and for the difference image $b = b_{obj} - b_{bgr}$ we thus have

$$\begin{aligned} \bar{b}[m,n] &= 0 \\ \sigma^2[m,n] &= \sigma^2_{obj}[m,n] + \sigma^2_{bgr}[m,n] \end{aligned} \tag{5.5}$$

Here we must assume $\sigma^2[m,n] = \sigma^2$ to be constant across the image area. In practice, this is achieved by means of level-dependent quantization (p. 125), variance smoothing (Section 7.1) or by using a sufficiently representative value of the variance across the entire image area.

To illustrate the concepts introduced, we continue by discussing a simple but fully realistic detection process.

5.2.1 Detection of astronomical objects

By far the majority of celestial objects are invisible to the naked eye. This is not exclusively due to their faintness, that is, the small number of photons reaching our eye. Part of the explanation is the presence of other photons, the noise from which 'drowns' such objects. This phenomenon is illustrated in Figure 5.2, which shows three CCD images of a certain sky area, recorded with the same telescope, but with increasing exposure times. The same effect is achieved by using a telescope with increasingly large apertures and/or more sensitive light detectors (if possible).

As the exposure time is increased by a factor of 16 between Figure 5.2c and Figure 5.2a, the former may well be regarded as the sum of 16 individual versions of Figure 5.2a – which is *not* the same as multiplying the image values by 16 (cf. Section A.5 of Appendix A). The total number of detected objects – galaxies and stars – obviously increases with exposure duration.

For CCD images of the sky, the following approximations are quite adequate: first, the sky background light is constant over the image; and second, the statistical variations are caused by photon noise (as opposed to electronic noise). Figure 5.3

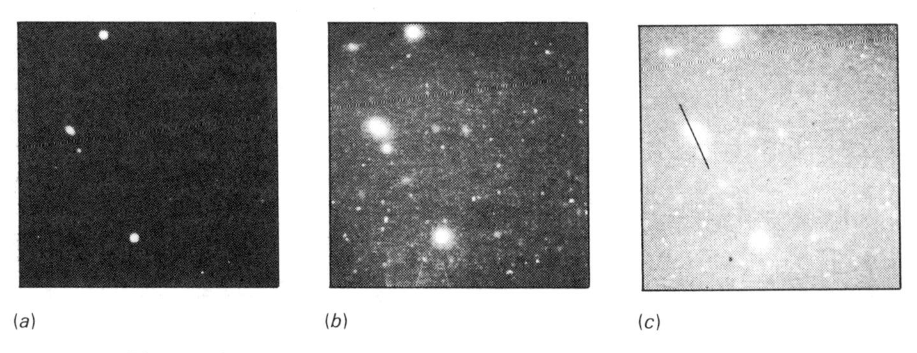

(a) (b) (c)

Figure 5.2 Exposure time (a) 5 min, (b) 40 min, and (c) 80 min.

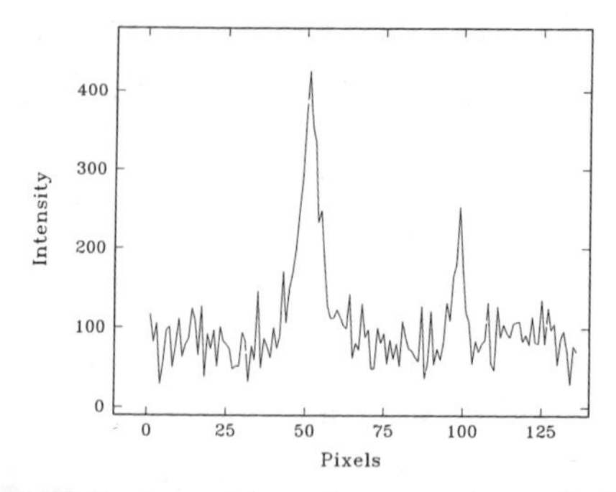

Figure 5.3 Intensities along section of CCD image.

shows the intensity values along the cut in Figure 5.2c. Both the constant background light with its noise and the superimposed faint objects are recognizable.

Figure 5.4 indicates the principle behind the detection process. For the sake of clarity, the violent intensity variations present in Figure 5.3 have not been reproduced; instead, they have been replaced symbolically by a shaded area of width 2σ. The intensity values are thus to be found 'mainly' in the interval $(b - \sigma, b + \sigma)$.

The intensity at a given point arises from two components: a 'signal' s and a 'background' b, the fluctuations of which are responsible for the noise σ. Expressing the intensities as photon numbers ensures that $\sigma = \sqrt{b}$. A 1σ detection occurs in those areas with $s \geqslant \sqrt{b}$. Note that $\sqrt{s + b}$ is not used as threshold value, the reason being that we are testing – with the statistical usage – the 'null hypothesis' $s = 0$.

In Figure 5.4a, no object is detected. If, next, the exposure time is increased by a factor a, the new signal s' and the new background b' become

$$s' = as \qquad \text{and} \qquad b' = ab$$

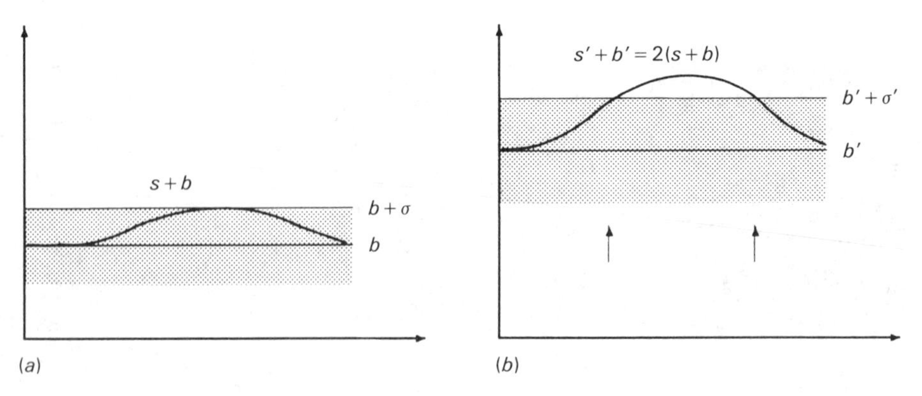

Figure 5.4 Astronomical object detection.

The new 1σ threshold is now located at $\sqrt{b'} = \sqrt{ab}$, and the detection condition $s' \geqslant \sqrt{b'}$ becomes $as \geqslant \sqrt{ab}$, that is,

$$s \geqslant \sqrt{b/a} \tag{5.6}$$

– a threshold reduced by the factor $\sqrt{a}$. One might also say that the signal/noise ratio $s/\sqrt{b}$ is increased by a factor $\sqrt{a}$. As shown in Figure 5.4b, one now detects an object located between the arrows.

As will be seen, the notion of an 'object' will depend – rather unsatisfactorily – on the detection resources, and a substantial part of the physical object may well fall below the detection limit.

EXAMPLE 5.9

We observe two identical astronomical objects, 1 and 2, 1 located at a distance r_1, 2 at a distance r_2. The ratio between the 'signals' (the amounts of energy) received from the two objects is

$$S_2/S_1 = (r_1/r_2)^2 \tag{5.7}$$

(cf. p. 18). The noise, which arises from the background light of the night sky, is the same for both objects.

In order to observe the two objects with the same signal/noise ratio and, consequently, with the same detection statistics, the ratio between the exposure times becomes (according to equations (5.7) and (5.6))

$$t_2/t_1 = (r_1/r_2)^4 \tag{5.8}$$

In practice, the theoretical inverse-square law (equation (5.7)) is thus replaced by this fourth-power law.

5.2.2 Detection by moments

As has been repeatedly stated, the level distribution for an image with background and object is characterized by two more or less well-defined components or 'peaks' (cf. Figures 4.6 and 4.9). Thus, the presence of an object might be established through a criterion testing the level distribution for this attribute. A simple and effective method is to evaluate a few of the distribution *moments*, which are expressions of the form

$$M_c = \sum n^c p_n$$

where, evidently, $M_0 = 1$, $M_1 = E(p)$, and $M_2 = V(p) + M_1^2$. Using a continuous probability density $f(x)$ for the representation of the level distribution, p becomes

$$M_c = \int x^c f(x)\, \mathrm{d}x$$

If, in an image area with N pixels, N_b background pixels and N_o object pixels are present, so that $N_b + N_o = N$, the level distribution (cf. Example 5.7) may be expressed as

$$f(x) = \frac{P}{\sqrt{2\pi}\sigma} e^{-(x-b)^2/2\sigma^2} + \frac{Q}{\sqrt{2\pi}\sigma'} e^{-(x-b-b')^2/2\sigma'^2}$$

where $P = N_b/N$ and $Q = N_o/N$ are the relative pixel numbers – or, if you prefer, the a priori probability of detection if an arbitrary pixel in the area is selected.

EXAMPLE 5.10
Show that, for this distribution, the mean value and variance are, respectively,

$$M_1 = b + Qb' \quad \text{and} \quad M_2 - M_1^2 = P\sigma^2 + Q\sigma'^2 + PQb'^2 \tag{5.9}$$

For a typical detection problem, b' is small, and Q is considerably smaller than P. Hence, M_1 is not likely to differ markedly from its value b in the case where no object was present. The term $Q\sigma'^2$ is, however, not necessarily negligible; in Figure 4.6, for instance, σ' is comparable to the level range shown. Thus, the variance will depend strongly upon the presence of an object in the distribution.

This trait is even more conspicuous for the higher-order moments, which may deviate strongly from the pure 'background values' and reveal the presence of an object.

It has been shown in this section that, in general, the same principles apply to detection of both a 'proper' (that is, connected) object, in a given image area, and extra photons in an individual pixel. Since the chosen detection threshold – which may possibly be pixel-dependent – causes a *segmentation* of the image into 'detected object' and 'background', our next task will be an examination of the object, that is, classification and characterization. Clearly, a positive recognition will almost always be interpreted as a detection which is 100 per cent certain.

5.3 Recognition

A recognition process involves a comparison between a given (unknown) object – which, thus, might be the result of thresholding and segmentation – and a number of reference objects. It is then decided whether these objects exhibit a sufficiently compelling resemblance to one of the reference objects to allow an identification. Alternatively, the object might be classified as 'unidentified'.

In order to quantify the resemblance between two objects (the unknown one and the reference) we return to the distance notion introduced in Section 2.5. The situation is indicated by the vectors in Figure 5.5. Here, we are to decide

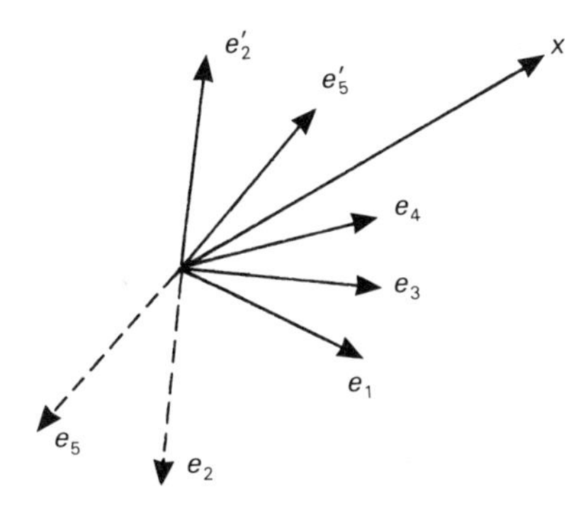

Figure 5.5 'Object' vector and 'reference' vectors.

whether the object vector x 'resembles' one or more among the reference vectors $e_1, e_2, \ldots, e_N$. First of all, we note from experience that neither the total intensity nor the norm of the vectors is of much use in the recognition process. Thus, it appears reasonable to standardize the reference vectors, and the natural choice is to assume them to be unit vectors.

Also, there will be no need to distinguish between the vectors e and $-e$ (positive and negative images); accordingly, all reference vectors will be chosen so as to ensure $x \cdot e \geqslant 0$, as shown in Figure 5.5, where, for example, e_2 has been replaced by $e'_2 = -e_2$.

The distance between object x and one of the reference vectors e is, then,

$$|x - e|^2 = |x|^2 + |e|^2 - x \cdot e - e \cdot x = |x|^2 + 1 - 2x \cdot e$$

for real images. The distance is thus minimal if $x \cdot e$ is maximal, and the classification amounts to assigning the image or vector x to the same class as the image e for which $x \cdot e$ is maximal.

The criterion for x to belong to the class defined by e_k is thus

$$|x \cdot e_k| \geqslant |x \cdot e_l| \qquad \text{for all } l \tag{5.10}$$

where the es are the original unit vectors irrespective of sign.

EXAMPLE 5.11

A three-dimensional space is divided into classes corresponding to minimal distance from the unit vectors

$$e_1 = (0, 1, 0) \qquad e_2 = (1, 1, 1)/\sqrt{3} \qquad e_3 = (2, 2, -1)/3$$

To which class does the vector $x = (1, 4, 2)$ belong?

According to this criterion, maximal distance between the reference vectors – that is, $e_k \cdot e_l = 0$ for $k \neq l$ – is desirable. We thus prefer the reference vectors to constitute an orthonormal set.

5.3.1 Recognition of simple objects

The general recognition principles above will now be made a little more specific. First, we substitute 'images' for 'vectors', and, to begin with, we restrict ourselves to very small images. Typically, the reference images will possess an extremely simple, idealized structure serving the purposes of identification of similar structures in more realistic images. For example, the simplest possible object is a single point, that is, a reference image of type

$$e_p = \begin{array}{|c|c|c|}\hline 0 & 0 & 0 \\ \hline 0 & 1 & 0 \\ \hline 0 & 0 & 0 \\ \hline\end{array}$$

where the surrounding pixels have been ignored. In the context of detection and recognition, the opposite of this object is the uniformly illuminated area

$$e_0 = \frac{1}{3}\begin{array}{|c|c|c|}\hline 1 & 1 & 1 \\ \hline 1 & 1 & 1 \\ \hline 1 & 1 & 1 \\ \hline\end{array}$$

where the factor $\frac{1}{3}$ ensures a normalization of this object. Since $e_p \cdot e_0 = \frac{1}{3}$, these two objects are not orthogonal; however, e_0 and $e_p - (e_p \cdot e_0)e_0$ are orthogonal (cf. p. 57), and the two new objects produce the same linear combinations as those of e_0 and e_p. Normalization leads to the reference objects

$$e_0 = \frac{1}{3}\begin{array}{|c|c|c|}\hline 1 & 1 & 1 \\ \hline 1 & 1 & 1 \\ \hline 1 & 1 & 1 \\ \hline\end{array} \quad \text{and} \quad e_1 = \frac{\sqrt{2}}{12}\begin{array}{|c|c|c|}\hline -1 & -1 & -1 \\ \hline -1 & 8 & -1 \\ \hline -1 & -1 & -1 \\ \hline\end{array}$$

These reference objects, specified as simple arrays, are called *masks* or *templates*.

EXAMPLE 5.12

On a radar screen, the following digitized object has been detected:

$$x = \begin{array}{|c|c|c|}\hline 38 & 7 & 0 \\ \hline 8 & 20 & 1 \\ \hline 5 & 4 & 8 \\ \hline\end{array}$$

Since $x \cdot e_0 = 30.33$ and $x \cdot e_1 = 10.49$, x must be classified as a 'non-point' according to the criterion discussed.

Another simple object is a *line*. Possible templates are

$e_1' =$

0	0	0
1	1	1
0	0	0

$e_2' =$

0	1	0
0	1	0
0	1	0

$e_3' =$

1	0	0
0	1	0
0	0	1

$e_4' =$

0	0	1
0	1	0
1	0	0

together with the above 'counter-example' e_0. Orthogonalization of e_0 and e_1' followed by normalization yields

$$e_1 = \frac{\sqrt{2}}{6} \begin{array}{|c|c|c|}\hline -1 & -1 & -1 \\ \hline 2 & 2 & 2 \\ \hline -1 & -1 & -1 \\ \hline \end{array}$$

This object and the analogous ones

$$e_2 = \frac{\sqrt{2}}{6} \begin{array}{|c|c|c|}\hline -1 & 2 & -1 \\ \hline -1 & 2 & -1 \\ \hline -1 & 2 & -1 \\ \hline \end{array} \qquad e_3 = \frac{\sqrt{2}}{6} \begin{array}{|c|c|c|}\hline 2 & -1 & -1 \\ \hline -1 & 2 & -1 \\ \hline -1 & -1 & 2 \\ \hline \end{array} \qquad e_4 = \frac{\sqrt{2}}{6} \begin{array}{|c|c|c|}\hline -1 & -1 & 2 \\ \hline -1 & 2 & -1 \\ \hline 2 & -1 & -1 \\ \hline \end{array}$$

are easily seen to constitute an orthonormal set. The set may be employed for identification of possible line structures in an object. In this very simple case, vertical and horizontal lines as well as diagonals are the preferred objects.

A further simple object is a *corner*. For identification of corners with vertical or horizontal edges, the following templates are adequate:

−1	1	1
−1	1	1
0	−1	−1

1	1	−1
1	1	−1
−1	−1	0

−1	−1	0
1	1	−1
1	1	−1

0	−1	−1
−1	1	1
−1	1	1

They are all orthogonal to e_0 but only approximately orthogonal to each other. Due to the simple structure of these templates, this drawback might be ignored; alternatively, the set may be orthogonalized (cf. Example 5.14).

Finally, there exist templates for the recognition of *edges*, that is, boundaries between two image areas with markedly different intensities. The following are typical templates for the detection of horizontal, vertical, and diagonal edges, respectively:

−1	−1	−1
0	0	0
1	1	1

−1	0	1
−1	0	1
−1	0	1

0	−1	−1
1	0	−1
1	1	0

1	1	0
1	0	−1
0	−1	−1

EXAMPLE 5.13 (Gram–Schmidt orthogonalization)

For $K+1$ linearly independent signals $f_1, f_2, \ldots, f_K$ and f, projection (cf. p. 58) leads to a linear combination $\bar{f} = \Sigma\, a_k f_k$ so that $f - \bar{f}$ is orthogonal to all the K first signals. Obviously, $f_1, \ldots, f_K$ and $f - \bar{f}$ have the same linear combinations as the original $K+1$ signals.

With the aid of this method, orthogonalize the set consisting of the templates e_0 as well as

$e_1' =$

0	1	0
1	0	1
0	1	0

and $e_2' =$

1	0	1
0	−1	0
1	0	1

EXAMPLE 5.14

Orthogonalize the set consisting of the four edge templates above and the neutral template e_0.

Table 5.2 Orthogonal templates for line and edge recognition.

neutral template ($\times 1/3$)

1	1	1
1	1	1
1	1	1

line templates ($\times 1/6$)

0	3	0
−3	0	−3
0	3	0

−3	0	3
0	0	0
3	0	−3

1	−2	1
−2	4	−2
1	−2	1

−2	1	−2
1	4	1
−2	1	−2

edge templates ($\times \sqrt{2}/4$)

1	$\sqrt{2}$	1
0	0	0
−1	$-\sqrt{2}$	−1

1	0	−1
$\sqrt{2}$	0	$-\sqrt{2}$
1	0	−1

0	−1	$\sqrt{2}$
1	0	−1
$-\sqrt{2}$	1	0

$\sqrt{2}$	−1	0
−1	0	1
0	1	$-\sqrt{2}$

(a)

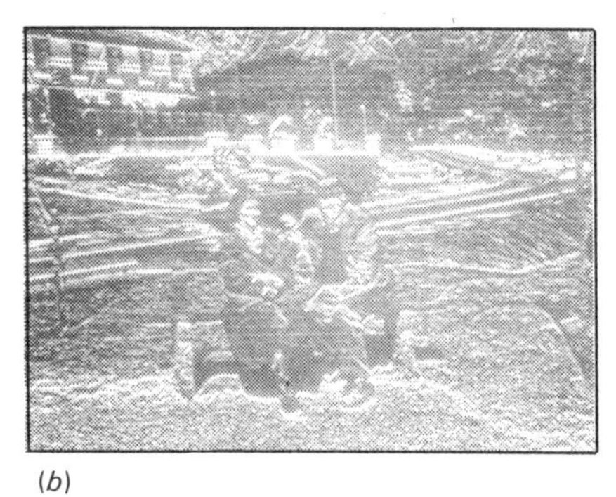
(b)

(c)

Figure 5.6 Line and edge recognition.

Thus, a multitude of templates are available for the recognition of certain types of object. In the case of 3×3 templates, nine linearly independent templates may be in play simultaneously and detect more than one type of structure. As an example of this procedure, a set of nine orthogonal templates, for simultaneous recognition of both lines and edges, is listed in Table 5.2 (see opposite).

Finally, Figure 5.6 shows an image (Figure 5.6a) in which a horizontal line template and a vertical edge template have been applied at each point. The result is a couple of new images (Figures 5.6b and 5.6c) displaying the corresponding scalar products. The method illustrated here leads to a new and very important notion of detection and recognition, to be discussed in the next section.

5.4 Correlation

Figure 5.7 shows a reference object e and a photograph b, known to represent the object somewhere. In the previous section, basic objects were identified by means of templates, movable at will throughout the image. Similarly, we now use the

$e[m,n] =$

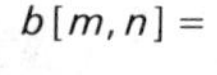

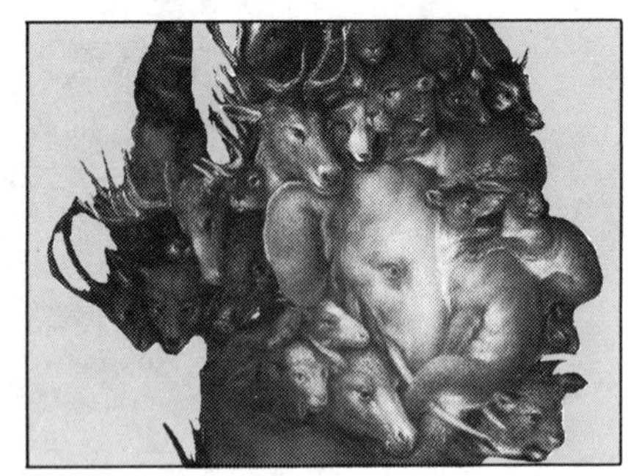

Figure 5.7 'Find the hidden animal'.

reference image e to perform a 'scan' of the observed image b, in the sense that scalar products are formed.

This procedure amounts to a *translation* of e (cf. equation (3.5)) to position $[k, l]$, where the scalar product

$$c = b \cdot T_{kl} e$$

is computed; T_{kl} denotes the translation in question. More explicitly, we construct

$$c = \sum_m \sum_n b[m,n] (T_{kl} e[m,n])^* = \sum_m \sum_n b[m,n]\, e[m-k, n-l]^*$$

The scalar product c now depends on k and l; it may thus be regarded as a digital image itself:

$$c[k,l] = \sum_m \sum_n e[m-k, n-l]\, b[m,n] \tag{5.11}$$

assuming the images to be real. The image c is called the *(cross)-correlation image* between e and b; Figures 5.6b and 5.6c provide an example.

In the analog case, the definition is as follows:

$$c(x,y) = \int_{y'} \int_{x'} e(x'-x, y'-y) b(x', y')\, \mathrm{d}x'\, \mathrm{d}y'$$

Figure 5.8 shows the correlation image between e and b from Figure 5.7.

For the scalar product $f \cdot g$ between two signals (images) f and g, the *Cauchy–Schwarz inequality* holds:

$$|f \cdot g| \leqslant |f|\,|g| \tag{5.12}$$

where equality obtains only if f and g are proportional. (The inequality is proved by utilizing $|f + ag|^2 \geqslant 0$ for the particular choice of the constant $a = -f \cdot g / |g|^2$.)

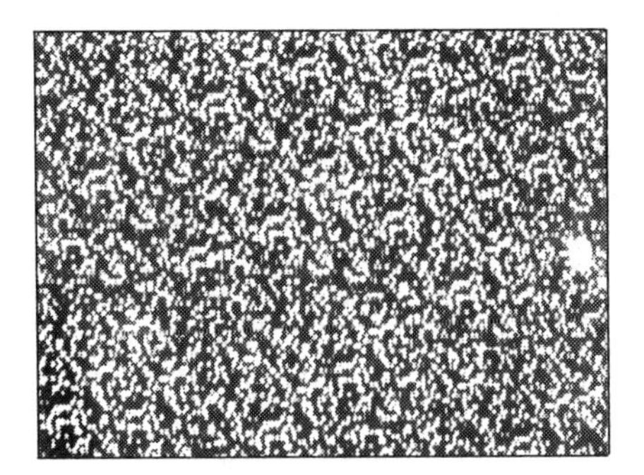

Figure 5.8 Localization of the hidden animal.

For a pixel $c[k,l] = b \cdot T_{kl}e$ in a correlation image, we have the upper limit

$$|c[k,l]| \leqslant |b||T_{kl}e| = |b||e| = |b| \tag{5.13}$$

if e is normalized. Here, $|c[k,l]| = |b|$ only holds if b and $T_{kl}e$ are proportional (in the present context, this means identical). An identification between b and $T_{kl}e$ is thus reflected by $|c[k,l]|$ being the maximal value in the image $|c[m,n]|$ consisting of the absolute values in the correlation image.

It might be stated that an *identification process* in the original image is replaced by a *detection process* in the correlation image. The relevant concept is that of area detection as described in the previous section. The maximal value $|b|$ obtained in the case of a recognition with a certainty of 100 per cent never occurs in practice. Nevertheless, it is of some significance for the determination of the relevant detection thresholds.

These thresholds are, as usual, related to the statistical properties of the image. For a given pixel $[k,l]$ in the correlation image, we have (for real images)

$$E(\mathrm{c}[k,l]) = \sum_{m,n} e[m-k, n-l]\, E(\mathrm{b}[m,n]) \tag{5.14}$$

$$V(\mathrm{c}[k,l]) = \sum_{m,n} e[m-k, n-l]^2 V(\mathrm{b}[m,n]) \tag{5.15}$$

where the symbols b and c serve as a reminder of the fact that the above quantities are, strictly speaking, stochastic variables associated with their respective pixels. In this context, the reference images are viewed as arrays of well-defined numbers, that is, as coefficients for the variables in question.

With the notation $E(\mathrm{b}[m,n]) = b[m,n]$ and $E(\mathrm{c}[m,n]) = c[m,n]$, equations (5.11) and (5.14) are identical.

EXAMPLE 5.15

If e is a 'binary template', that is, if e can only assume the value 0 or 1, and if b is Poisson distributed in all pixels, we obtain from the equations (5.14) and (5.15) the following expression:

$$V(\mathrm{c}[k,l]) = E(\mathrm{c}[k,l]) = c[k,l]$$

In this case – where e has not been normalized – $\mathrm{c}[k,l]$ is a sum of Poisson variables and is therefore itself Poisson, a result which is not generally true. If, however, $\mathrm{b}[m,n]$ is approximated with a normal distribution (in each pixel), then $\mathrm{c}[m]$ is also approximately normally distributed.

Assuming as usual that $V(\mathrm{b}[m,n])$ is a constant σ^2 across the image, one has

$$V(\mathrm{c}[k,l]) = \sigma^2 \sum e[m-k, n-l]^2 = \sigma^2 \,|\, e \,|^2 = \sigma^2$$

if the reference image e is normalized. The correlation image variance will thus be equal to that of the original image (and constant as well), and the detection principles discussed can be applied.

With this formulation of the problem, we have to test the hypothesis of the existence of a correlation against the alternative of 'no correlation'. In the latter case, the correlation image must be roughly constant, equal to the average

$$\bar{c} = \frac{1}{MN} \sum_{k,l} c[k,l] = \frac{1}{MN} \sum_{k,l} \sum_{m,n} e[m-k, n-l]\, b[m,n]$$

over the image – which, normally, should not be confused with the statistical mean value for the individual pixels. Further,

$$\bar{c} = \frac{1}{MN} \sum_{m,n} b[m,n] \sum_{k,l} e[m-k, n-l] = \left(\sum e\right) \bar{b} \tag{5.16}$$

where $\bar{b}$ is the average of b, and Σe is the sum of the values of the es.

EXAMPLE 5.16

Show that, if e appears in an orthonormal set together with the constant (neutral) template, then $\bar{c} = 0$.

EXAMPLE 5.17

A template e is given by

$$e = \frac{1}{\sqrt{20}} \begin{array}{|c|c|c|} \hline 0 & -1 & 0 \\ \hline -1 & 4 & -1 \\ \hline 0 & -1 & 0 \\ \hline \end{array}$$

This template is correlated with the image b, having a representative variance $\sigma^2 = 100$, and a segment of the resulting correlation image is as follows:

−5.4	−12.2	7.0	8.7	13.2	2.4
10.9	1.4	11.6	−5.1	0.8	3.1
−7.1	4.4	9.6	6.1	4.7	−8.2
−9.1	0.3	15.7	12.5	11.0	5.9
1.6	13.0	16.7	14.2	11.0	4.7
−8.8	8.9	12.3	14.9	9.5	3.0

Apart from the strongly simplified problem, we have a '1σ correlation', since the shaded pixels fall above the detection limit $\bar{c} + \sqrt{100} = 10$. The values 10.9, 11.6, and 13.2 are not found in neighbouring pixels and are regarded as fluctuations.

Clearly, if one is satisfied with a loose detection criterion as in Example 5.17, it should be taken into account that detections in neighbouring pixels *might* be accidental.

Finally, we summarize the results of this section:

The existence of a correlation between a normalized reference image e and an image b is quantified through a detection in the correlation image

$$c[k, l] = b \cdot T_{kl} e$$

The absence of correlation is characterized by a correlation image the pixels of which have the same mean value $\bar{c} = (\Sigma\ e)b$ and a variance σ^2 equal to that of the original image.

5.5 Correlation and spectral estimation

Because of the formal similarity between correlation and convolution, it seems natural to perform the practical evaluation of a correlation image by means of a digital Fourier transform. To simplify the notation, we consider Fourier transforms of one-dimensional analog signals for the time being. Let f and g denote two such – possibly complex – signals. Their correlation signal is

$$c(t) = \int f(s)g(s-t)^* \, \mathrm{d}s$$

Note that, for real and even signals, correlation and convolution are identical.

The Fourier transform of c becomes

$$C(\omega) = \int c(t)\mathrm{e}^{-\mathrm{i}\omega t} \, \mathrm{d}t = \int\int f(s)g(s-t)^* \mathrm{e}^{-\mathrm{i}\omega t} \, \mathrm{d}s \, \mathrm{d}t$$

$$= \int_s f(s)\mathrm{e}^{-\mathrm{i}\omega s} \int_t (g(s-t)\mathrm{e}^{-\mathrm{i}\omega(s-t)})^* \, \mathrm{d}t \, \mathrm{d}s$$

$$= F(\omega)G(\omega)^* \tag{5.17}$$

where F and G are the Fourier transforms of f and g, respectively.

The cross-correlation signal between two signals may thus be obtained by Fourier transformation of each signal, complex conjugation of one of these, multiplication and inverse Fourier transformation.

In particular, f and g may be the same signal, in which case we are looking for similarities between different parts of the signal – or different image areas. If so, the *autocorrelation signal* (or *image*) for f is

$$\phi(t) = \int f(s)f(s-t)^* \, \mathrm{d}s \tag{5.18}$$

Here, according to equation (5.17),

$$\Phi(\omega) = |F(\omega)|^2 \tag{5.19}$$

where Φ is the Fourier transform of ϕ. We have thus dug out the *energy spectrum* Φ for f, and equation (5.19) may be expressed as follows:

> The energy spectrum for a signal is the Fourier transform of its autocorrelation signal.

For images, this quantity specifies the distribution of the (electromagnetic) energy over the *spatial* frequencies – not to be confused with the temporal frequencies (cf. equation (1.11)) or wavelengths.

The one-dimensional autocorrelation signal $\phi(t)$ is an *even* function of t, and the value $\phi(0)$ is a maximum (cf. expression (5.13)):

$$\phi(0) = \int |f(t)|^2 \, \mathrm{d}t = |f|^2$$

Both for convolution and correlation, edge effects will be present. If, for instance, a 1000×500 image is convolved or correlated with a 4×3 point spread function or template, the result is a 1003×502 image. Here, the image created will be of roughly the same size as the original – as expected, since PSFs and templates are normally 'small' images. In the case of an autocorrelation, however, the situation is quite different. The scalar product appearing in the definition of ϕ must be summed over a number of pixels varying from that of the original image (that is, for $\phi[0,0]$) all the way down to 1. Clearly, the autocorrelation image $\phi[m,n]$ will be most useful for small values of m and n.

In all circumstances, one should use the average value of the scalar product (pp. 54–5) in order to compare the values of ϕ. From now on, we shall therefore employ the definition

$$c[m] = \frac{1}{N} \sum f[n]\, g[n-m] \tag{5.20}$$

for the cross-correlation signal c of two real signals f and g; this definition also applies to the autocorrelation signal ϕ, for which $f = g$. The integer N denotes the number of terms in the sum.

This procedure is likewise formally convenient if, in order to avoid edge effects, one envisages the image to be duplicated outside its definition area, as shown in Figure 5.9.

EXAMPLE 5.18

Find the autocorrelation image for the following 4×3 image

1	0	1	0
0	1	0	1
1	0	1	0

(a) with and (b) without periodic duplications.

Figure 5.9 Periodically repeated image.

EXAMPLE 5.19
Find the autocorrelation images (without duplications) for

(a) $b(x, y) = x + y, \quad x \in (-3, 3), y \in (-3, 3)$

(b) $b[m, n] = mn, \quad m \in [0, 5], n \in [0, 4]$

In an autocorrelation process, the image acts as its own template. This means that the autocorrelation image detects repetitions in the original image. If, in particular, *periodicities* in the original image are present, the autocorrelation image will contain 'unusually large' values. Thus, the method ought to be suited for the detection of a periodic content in a signal or an image. This feature is illustrated by computing the autocorrelation for a periodic signal $f(t), t \in (0, T)$, with the Fourier series $f(t) = \Sigma\, a_n e^{in\omega_0 t}$ (where $\omega_0 = 2\pi/T$). Here,

$$\phi(s) = \frac{1}{T}\int_0^T \sum_{n=-\infty}^{\infty} a_n e^{in\omega_0 t}\left(\sum_{m=-\infty}^{\infty} a_m e^{im\omega_0(t-s)}\right)^* dt$$

$$= \frac{1}{T}\sum_{m,n} a_n a_m^* e^{im\omega_0 s}\int_0^T e^{i(n-m)\omega_0 t}\, dt$$

$$= \sum_{n=-\infty}^{\infty} |a_n|^2 e^{in\omega_0 s}$$

since the harmonic basis signals are orthogonal. The autocorrelation signal is thus periodic and has the same period as the signal itself; the coefficients in its Fourier series are the squared moduli of the original coefficients. The factor $1/T$ reflects the new definition of the correlation signal (cf. p. 55).

The detection of periodicities in a signal or image and the evaluation of their amplitudes by means of autocorrelation may seem clumsy, since the methods in Sections 2.7 and 3.3 were introduced precisely to this end. Left unsaid, however, was the fact that the simple Fourier analysis is rather sensitive to noise. As is obvious from Figure 5.3, any signal – whether periodic or not – is undetectable if its maximum value is less than the noise amplitude (called σ in Figure 5.3). In the autocorrelation signal, however, the repetitions due to a periodic signal will cause a systematic amplification, whereas the noise will normally be smoothed away.

For images or signals where the statistical nature is to be highlighted, one avoids the term 'Fourier analysis'; instead, the more prudent term *spectral estimation* is employed.

In the above considerations concerning cross-correlation, only one of the signals to be correlated was regarded as stochastic, and the cross-correlation signal was, hence, a *linear combination* of stochastic variables. (If one wishes to stress the contrast with 'stochastic', the term *deterministic* is used for a 'normal' signal.) In general, the linear combination is replaced by a *sum of products*. For instance, the autocorrelation signal for the stochastic signal f is given by

$$\phi[m] = \frac{1}{N} \sum_n \mathsf{f}[n]\mathsf{f}[n-m]$$

$$= \frac{1}{N} \mathsf{f} \cdot (T_m \mathsf{f}) \qquad (5.21)$$

As ϕ is considered to be a deterministic signal, this expression calls for some further explanation. Since ϕ is written as an average over digital time, it is redefined in the stochastic case:

$$\phi[m] = E(\mathsf{f} \cdot (T_m \mathsf{f})) \qquad (5.22)$$

Here it has been assumed that averages and statistical means are identical. As mentioned, this assumption does not necessarily hold automatically. Nevertheless, it is justified in many cases of practical interest; if so, f is said to be an *ergodic* stochastic signal.

We illustrate these considerations by providing an example of the extraction of a 'meaningful' signal from a noisy background. Let us examine a one-dimensional signal $\mathsf{f}[n]$, considered to be a section through an image as shown in Figure 5.3. We assume that f is a sum of a signal s and a background b, s being a deterministic signal and b a stochastic one. The background b is assumed to be of constant mean value b and variance σ^2, that is, independent of digital time n. Since the constant value b may be absorbed in s, we further assume $b = 0$.

The autocorrelation signal ϕ_f for f is now

$$E(\mathsf{f}[n]\mathsf{f}[n-m]) = E((s[n] + \mathsf{b}[n])(s[n-m] + \mathsf{b}[n-m]))$$

hence, it is a sum of four terms:

$$\phi_f[m] = \phi_s[m] + E(s[n]\mathsf{b}[n-m]) + E(s[n-m]\mathsf{b}[n]) + \phi_b[m]$$

with similar symbols for the autocorrelation signals ϕ_s and ϕ_b for s and b, respectively. Since $E(\mathsf{b}) = 0$, we have

$$\phi_f = \phi_s + \phi_b$$

In this equation, $\phi_b[m] = E(\mathsf{b}[n]\mathsf{b}[n-m])$, and if the stochastic variables associated with the various digital times (pixels) are independent, then

$$\phi_b[m] = \begin{cases} E(\mathsf{b}^2) & \text{for } m = 0 \\ 0 & \text{otherwise} \end{cases}$$

so that ϕ_b will be proportional to a delta signal. Finally, $E(\mathsf{b}^2) = \sigma^2$, since $E(\mathsf{b}) = 0$. With the assumptions stated, we thus have

$$\phi_b[m] = \sigma^2 \delta[m] \qquad (5.23)$$

A Fourier transformation of this equation shows b to be of constant power spectrum

$$\Phi_b[n] = \sigma^2 \tag{5.24}$$

A stochastic signal of constant power spectrum is called *white noise*. White noise appears in many theoretical developments and in many cases of practical interest as a convenient description of the spectral properties of a noisy signal, even if the set of assumptions differs from that given above.

In conclusion, the power spectrum Φ_f for f consists of two components,

$$\Phi_f = \Phi_s + \sigma^2$$

and the quantity Φ_s may be estimated by performing a DFT of the samples of f, if σ^2 is known.

5.6 Image moments

When characterizing images and their objects, *image parameters* are used. If, for instance, an 'image' consists of one pixel only, in which the photon count obeys a Poisson distribution, the distribution parameter will provide a typical example. More realistic parameters are associated with shape, intensity distribution, and so on, for certain objects or segments in the image.

We have already described how objects may be extracted from the image background by segmentation, possibly after background subtraction. For simplicity, we shall assume in the present section that the images to be investigated contain a single, isolated object only, the parameters of which we are to estimate. Moreover, we assume that the statistical properties of the image in question – which normally has undergone some kind of processing – are well defined. It turns out to be convenient to adopt a slightly different point of view, where the image (the object) $b(x, y)$ is regarded as a two-dimensional probability density for the variable (x, y). We assume the quantity $b_{00} = \int\int b(x, y)\, dx\, dy$ to be unity. If this is not the case, the image $b(x, y)/b_{00}$ is considered instead.

The most important parameters for an image (object) are its *moments*, that is, the quantities

$$b_{kl} = \int\int x^k y^l b(x, y)\, dx\, dy \tag{5.25}$$

This expression is one of the $k + l + 1$ *moments of order* $k + l$ *for* b. By far the most important of these are the moments of order 0, 1, and 2:

$$b_{00} = \int\int b(x, y)\, dx\, dy$$

$$b_{10} = \int\int xb(x, y)\,dx\,dy \qquad b_{01} = \int\int yb(x, y)\,dx\,dy$$

$$b_{20} = \int\int x^2b(x, y)\,dx\,dy \qquad b_{02} = \int\int y^2b(x, y)\,dx\,dy \tag{5.26}$$

$$b_{11} = \int\int xyb(x, y)\,dx\,dy$$

As mentioned, the *total intensity* b_{00} of the image equals 1. Its *centre of gravity* or, simply, *centre* has coordinates (b_{10}, b_{01}). If, then, we examine the two-dimensional variable $(\mathsf{x} - b_{10}, \mathsf{y} - b_{01})$, its distribution becomes *centred*, since both its first-order moments are zero. This is equivalent to the introduction of a new image coordinate system with its origin at the centre (b_{10}, b_{01}) – or, alternatively, to translating the image by T_{xy}, with $(x, y) = (-b_{10}, -b_{01})$.

In this coordinate system, we may assume that

$$b_{00} = 1 \qquad \text{and} \qquad b_{10} = b_{01} = 0$$

With this choice, the *central* moments of second order,

$$b_{20}^c = b_{20} - b_{10}^2, \qquad b_{11}^c = b_{11} - b_{10}b_{01}, \qquad \text{and} \qquad b_{02}^c = b_{02} - b_{01}^2 \tag{5.27}$$

are identical with the usual moments b_{20}, b_{11}, and b_{02}. These quantities measure the extension and orientation of the object, and they are independent of – corrected for, so to speak – the position in the image.

EXAMPLE 5.20

Find the second-order moments (both non-central and central) for each of the two objects in the image below.

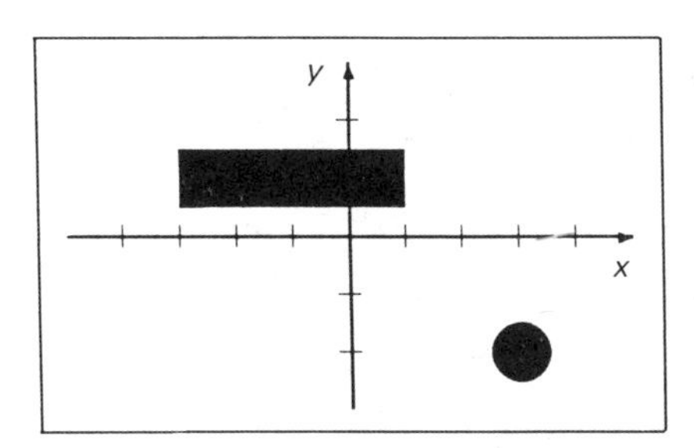

One object is a rectangle with sides of length 4 and 1, the other a circle centre $(3, -2)$ and diameter 1. The level is 1 across the rectangle and 3 across the circle.

It is possible to customize the three moments of second order further, correcting for *orientation* as well. Stated differently, the moment b_{20} characterizes the extension in the x-direction, and unless this direction bears some kind of relation to the object itself, b_{20} is not particularly interesting. Figure 5.10 shows an elliptical object positioned centrally in the coordinate system. It seems obvious that the moments calculated relative to the broken-line coordinate system yield a better characterization of the object than those calculated in the original system. If the original coordinate system is rotated through an angle θ, then the coordinates of a given image point change from (x, y) to (x', y'):

$$x' = x \cos\theta + y \sin\theta \qquad y' = -x \sin\theta + y \cos\theta \tag{5.28}$$

(cf. Example 3.7 and Chapter 6). Choosing the particular value θ corresponding to

$$\tan 2\theta = \frac{2b_{11}}{b_{20} - b_{02}}, \qquad \theta \in \begin{cases} \left(0, \dfrac{\pi}{2}\right) & \text{for } b_{11} > 0 \\ \left(\dfrac{\pi}{2}, \pi\right) & \text{for } b_{11} < 0 \end{cases} \tag{5.29}$$

the following situation occurs for the second-order moments in the new coordinate system:

$$b'_{20} = \text{maximal} \qquad b'_{11} = 0 \qquad b'_{02} = \text{minimal}$$

where 'maximal' and 'minimal' mean 'when θ varies'. The quantities in question are

$$\left.\begin{matrix} b'_{20} \\ b'_{02} \end{matrix}\right\} = \tfrac{1}{2}\left(b_{20} + b_{02} \pm \sqrt{(b_{20} - b_{02})^2 + 4b_{11}^2}\right) \tag{5.30}$$

The moments appearing on the right-hand side should be central.

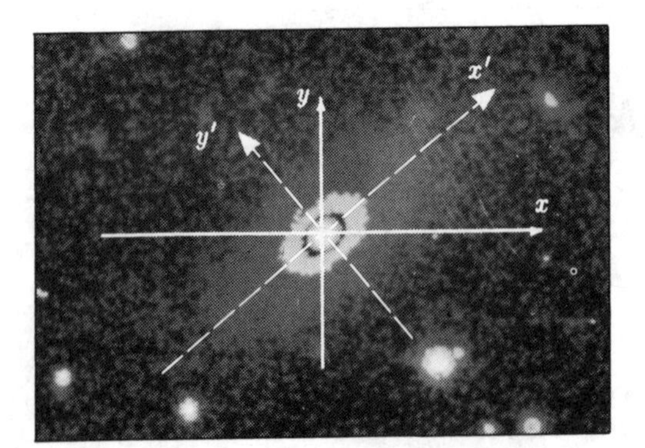

b_{20}	b_{11}	b_{02}
6.57	3.62	5.29

Figure 5.10 An elliptical galaxy and its central moments.

EXAMPLE 5.21
Prove the above results.

The two moments above characterize the object in itself in the sense that they have been corrected for position and orientation. Accordingly, they are termed *invariant moments*. For objects with elliptical symmetry, these moments are proportional to the squares of the ellipse axes. The above considerations may be generalized to invariant moments of third and higher orders as well.

EXAMPLE 5.22
Find the two invariant moments of second order for the elliptical galaxy in Figure 5.10.

EXAMPLE 5.23
Find the three central moments of second order for a rectangle with sides 4 and 1, rotated through $+60^\circ$ relative to its 'invariant' coordinate system.

Moments, particularly invariant moments, are powerful tools for object recognition and classification. For an image containing a number of objects, this process might consist in an evaluation of N moments or similar image parameters for all the objects. Each object thus gives rise to a point in an N-dimensional parameter space (called the *feature space*), and any similarity between two objects will be quantified as a small distance (Section 2.5) in this space. The recognition or classification of an object, according to this principle, will thus amount to a decision concerning the distance of the object point, in the feature space, from *either* a point defined by a reference object *or* a class of mutually nearby points, a so-called *cluster*.

It should be noted that the statistical properties of the original image are reflected in this space and the distances encountered here. An image parameter, for instance an invariant moment, will not cause the appearance of a *point* in the feature space – rather, it will cause a *probability density*. The mean value of this probability density will normally equal the parameter to be estimated, but a proper specification of an image parameter should include the variance as well. These variances do, of course, play a decisive role in the classification of the object.

EXAMPLE 5.24

An image contains seven objects for which the following moments are given:

Object no.	b_{10}	b_{01}	b_{20}	b_{11}	b_{02}
1	−5.0	−7.5	30.47	38.35	60.88
2	4.1	3.7	21.34	16.12	20.66
3	−3.3	8.2	16.40	−27.69	73.73
4	7.2	9.7	58.32	72.84	99.51
5	−9.9	3.5	104.95	−37.74	15.91
6	5.2	−3.2	34.46	−15.41	12.62
7	−6.5	−4.3	47.75	30.85	23.99

Find the two invariant second-order moments for all seven objects and plot them as points in a (b'_{20}, b'_{02})-plane.

It is next assumed that both b'_{20} and b'_{02} are normally distributed with the above values as statistical means, while the standard deviations for all the objects are $\sigma'_{20} = 1$ and $\sigma'_{02} = 1$. Objects 1, 2, and 3 are apples, while 4, 5, and 6 are pears. What is object 7?

EXAMPLE 5.25

An image is described by a background level of 100 and a standard deviation of 30. A portion of the image is shown:

83	112	117	101	119	106	137	127	113	97	95	87	106	91	102
121	131	109	120	107	134	115	112	117	122	73	109	120	96	90
117	149	140	144	135	139	131	126	147	136	109	115	99	98	103
115	102	138	148	166	169	178	160	137	130	146	134	118	103	97
121	139	144	132	182	181	207	207	174	153	151	127	128	118	107
128	135	136	173	193	217	272	344	255	178	175	129	141	131	94
125	145	151	154	193	293	421	485	399	238	193	130	125	118	121
121	125	143	167	210	269	387	509	432	269	180	129	108	95	132
135	118	145	165	194	225	291	338	331	262	187	171	111	127	140
124	121	115	113	136	143	183	195	200	156	159	139	128	123	137
126	97	119	120	115	143	149	146	156	139	149	124	116	145	110
75	95	122	118	104	128	124	129	122	150	131	122	116	125	138
99	100	92	132	130	132	126	120	118	106	135	98	103	134	109
107	104	130	94	107	97	115	108	111	95	114	118	96	101	128
100	125	118	126	102	115	130	124	96	94	104	95	132	115	76

Find the mean and standard deviation of the total intensity of the object extracted from the above portion by background subtraction and application of the detection thresholds (a) 1σ, (b) 2σ.

Hint: Use equations (5.5), disregarding the statistical fluctuations of the object itself.

5.6.1 Maximum likelihood estimation of total intensities

In our discussion of the recognition process, we said that the total intensity of an object would, in general, be unsuitable for classification purposes. If, however, an object has been recognized, that is, identified as being proportional to a reference object, this identification may be used for an estimation of the relevant constant of proportionality, characterizing the object completely. This principle may be useful for the estimation of total intensity, with which subject we conclude this chapter.

Thus, an object (vector) x is envisaged to have been identified as belonging to the class of vectors proportional to the reference vector e. If the latter has components $e[n]$, the recognized object must be of form $\bar{x}[n] = ae[n]$; however, due to the statistical fluctuations, we instead observe the components $x[n]$. We now determine the maximum likelihood (ML) estimator $\hat{a}$ for a, with the usual assumptions about its statistical properties.

First, if x is assumed to be a Poisson distributed stochastic signal, the probability of observing the value $x[n]$ is given by

$$p_{x[n]} = e^{-(ae[n])} \frac{(ae[n])^{x[n]}}{x[n]!}$$

and the probability of having obtained exactly the values observed is, thus, $\mathscr{L} = \prod_n p_{x[n]}$, with the corresponding logarithmic likelihood function

$$L(a) = \ln \mathscr{L} = \sum_n (x[n]\ln(ae[n]) - ae[n] - \ln(x[n]!))$$

The ML estimator of a is the value of a which maximizes L. Differentiation with respect to a shows this to be

$$\hat{a}^{\text{Poisson}} = \sum_n x[n] \Big/ \sum_n e[n] \tag{5.31}$$

Assuming, next, that $x[n]$ is normally distributed with mean $ae[n]$ and constant variance $\sigma^2[n] = \sigma^2$, the likelihood function becomes

$$L = -\sum_n (x[n] - ae[n])^2/2\sigma^2$$

which is a maximum when the signal distance $|x - ae|$ is a minimum, that is, when *ae* is obtained from x by *projection*. In this case, the ML estimator becomes (cf. equation (2.27))

$$\hat{a}^{\text{Gauss}} = e \cdot x = \sum_n e[n]\, x[n] \tag{5.32}$$

In particular, the total intensity (zeroth-order moment) x_{00} for the identified object may be found from one of the equations (5.31) or (5.32), where

$$x_{00}^{\text{Poisson}} = \sum_n x[n] \qquad \text{and} \qquad x_{00}^{\text{Gauss}} = \left(\sum_n e[n]\right) \sum_n e[n]\, x[n] \tag{5.33}$$

Note that the Poisson estimator does not use the prior knowledge of *e*.

EXAMPLE 5.26

The two images below represent, respectively, a reference object e and another object x, identified as being proportional to e:

$e[m, n] = \frac{1}{100}$

0.3	1.9	4.3	3.6	1.1
1.7	10.7	24.5	20.6	6.4
3.6	22.6	51.8	43.7	13.6
2.8	17.6	40.3	34.0	10.6
0.8	5.0	11.6	9.8	3.0

$x[m, n] =$

2	6	12	2	4
2	20	59	52	20
11	57	141	121	29
4	54	110	87	33
2	11	34	30	10

Estimate (statistically!) the total intensity and its standard deviation, if the image statistics is assumed to be described by (a) a Poisson distribution, (b) a normal distribution with standard deviation 6. Compare with the result obtained in the latter case with no knowledge of the reference object.

6

Geometric Operations

6.1 Overview

Among the image transformations occurring in practice, the purely geometrical ones are, of course, quite common: translations (cf. p. 73), rotations, magnifications, point and line symmetries, and so on. Other transformations alter the individual pixel values according to certain precepts independent of the image as such. These *level operations* will be discussed in Chapter 7. For now, we proceed with the former type – the *geometric operations*. Transformations in general may naturally be regarded as composites of these two types of operation.

The significance of geometric operations has already been mentioned in the context of 'information reduction' (data compression) – cf. p. 71. If, in particular, an image is to be stored for reference purposes, one would be in trouble if all kinds of translated, rotated, or enlarged versions were necessary – one standard version ought to suffice.

This remark is especially relevant as regards *recognition*; here, geometric operations play an essential role. Previously, we correlated an observed object and a reference object by submitting one object to a pure translation with respect to the other. Now, we extend the scope by invoking the more general class of geometric operations.

Due to the practical limitation arising from the fact that any digital image is endowed with a specialized geometric structure arising from its rectangular pixel grid, we assume for the moment that all images are *analog*. This simplification will be remedied in Section 6.5, which is devoted to *interpolation*; this notation will be viewed as a means of restoring a digital image in a new pixel grid. This is accomplished through an image manipulation of the form digital → analog → digital.

In a geometric operation, an image $b'(x, y)$ is altered into a new one, $b(x, y)$, consisting of the *same* intensity values, but placed in new positions:

$$b(x, y) = b'(x', y') \tag{6.1}$$

where the compact notation x' and y' masks the *coordinate transformation*

$$x' = x'(x, y), \qquad y' = y'(x, y) \tag{6.2}$$

If, for instance, b is the result of translating b' 2 units in the x-direction and 3 units in the y-direction (cf. Figure 3.4), then

$$b(x, y) = b'(x - 2, y - 3)$$

so that the relevant coordinate transformation is $x' = x - 2$, $y' = y - 3$.

In each case it should be stated clearly whether the operation is envisaged as acting upon the object or upon the coordinate system in which the object has been described. A common cause of confusion is the fact that the result of a geometric operation upon an object is identical to that obtained by exposing the coordinate system – regarded as a geometric object – to the *inverse* transformation.

EXAMPLE 6.1

An image b' is given by

$$b'(x, y) = 10\mathrm{e}^{-(x^2 + y^2/2)}$$

It is rotated 120° (anticlockwise) around the origin, resulting in the image b. What is the value $b(0.5, 1.3)$?

Solution. The coordinate transformation is, from Example 3.7,

$$x' = x \cos 120^\circ + y \sin 120^\circ \qquad y' = -x \sin 120^\circ + y \cos 120^\circ$$

Thus, the value $(x, y) = (0.5, 1.3)$ is associated with $(x', y') = (0.88, -1.08)$, having pixel value $b'(0.88, -1.08) = b(0.5, 1.3) = 2.58$. What is $b(1, 0)$?

The traditional usage for geometric operations is, unfortunately, somewhat unclear. A coordinate transformation as given in equations (6.2) is not necessarily a transformation in the sense of Section 3.1, since it is not required to conserve linear combinations of two-component vectors (x, y). Nevertheless, it will always *generate* an image transformation: it does not matter whether linear combinations are formed before or after the geometric operation has taken effect on the images in question. On the other hand, many interesting image transformations are not generated by coordinate transformations.

To avoid confusion, we shall use the mathematical term *map* for the new concept, as formulated in equations (6.2). The maps possessing the transformation property in the previous sense – where the signals, however, have only two components – will be studied in more detail in the next section.

(a) Original image

(b) Corrected image

Figure 6.1 Geometric decalibration (illustration of principle!).

A typical map is the geometric distortion introduced by the imaging apparatus. Such a distortion may be corrected in various ways. For instance, a set of reference points – points of known coordinates – may be available for the image. When the distortion and the subsequent correcting map are applied, the reference points should remain fixed, and the processing of the rest of the image must satisfy this constraint. The method is known as *geometric decalibration*.

For instance, the reference points might form a rectangular grid, to be mapped together with the image points. From the deformation of this line system, the image may be 'straightened out', as shown in Figure 6.1. The image distortions exemplified by this and similar maps have no general properties to be utilized in the decalibration process – except their being continuous. However, many realistic geometric operations obey very simple rules. Quite often, a certain image property is *conserved*, that is, present both before and after the operation. The following sections are devoted to such maps.

6.2 Affine maps

The most important geometric maps are those *preserving straight lines*: any line which is straight *before* being mapped will also be straight *afterwards*. This means that if a line l is specified by the *parametrization*

$$l:\ x = x_0 + te, \qquad t \in (-\infty, \infty) \tag{6.3}$$

then its points x are mapped onto points x' on a straight line too, say $x_0' + t'e'$. Here, x_0 is an arbitrary point (vector) on the line and e a vector specifying its direction.

A *linear map* is a map preserving linear combinations. This definition is exactly the same as the one given in Section 3.1; in the present context, it is specifically applied to vectors (signals with two components). If the points x on the line l are subjected to a linear map L, then $x' = L(x) = L(x_0) + tL(e)$, and the points x' will thus be found on the line l' determined by $x_0' = L(x_0)$ and in the direction $e' = L(e)$. A linear map is thus line-preserving.

The simple *translations* share this property: If the line l in equation (6.3) is translated by the vector a, it becomes

$$l': x' = (x_0 + a) + te, \qquad t \in (-\infty, \infty)$$

that is, a line with the same direction as l.

Other types of line-preserving maps exist. A very important class, the so-called *projective* maps, will be discussed in Section 6.4.

The property of linearity is easily transferred from the general signal transformations to the far simpler situation where only two signal components are present, x_1 and x_2 (see p. 78). In this case, only two basis vectors ('delta signals') $e_1 = (1, 0)$ and $e_2 = (0, 1)$ are necessary, and the vector $x = (x_1, x_2)$ may be written as the linear combination $x_1 e_1 + x_2 e_2$. Through the action of a linear map L, this vector changes to

$$x' = (x_1', x_2') = x_1' e_1 + x_2' e_2 = x_1 L(e_1) + x_2 L(e_2)$$

If, then, $L(e_1) = a_{11} e_1 + a_{21} e_2$ and $L(e_2) = a_{12} e_1 + a_{22} e_2$, the result is

$$x_1' = a_{11} x_1 + a_{12} x_2 \qquad x_2' = a_{21} x_1 + a_{22} x_2 \tag{6.4}$$

which is the explicit form of the most general *linear* map.

EXAMPLE 6.2

Show that this map is *invertible*, that is, the equation system (6.4) can be solved with respect to x_1 and x_2, unless the determinant $d = a_{11} a_{22} - a_{12} a_{21}$ vanishes. Also show that if (and only if) the determinant is 0, a whole line exists, the vectors of which are all mapped onto the zero vector.

Note that the translations T_a given by $x' = a + x$, that is,

$$x_1' = a_1 + x_1 \qquad x_2' = a_2 + x_2 \tag{6.5}$$

are not linear. Since they occur very frequently, however, linear maps, translations, and their composites are referred to collectively as *affine maps*.

Thus, the most general affine map is of the form

$$x_1' = a_1 + a_{11} x_1 + a_{12} x_2 \qquad x_2' = a_2 + a_{21} x_1 + a_{22} x_2 \tag{6.6}$$

and is composed of a linear map L and a translation T_a:

$$A = T_a L$$

The order of composition is significant, even if $B = L T_a$ is also an affine map.

The decomposition $A = T_a L$, where an affine map A is written as a composite of a linear map and a translation, is unique, since

$$A(0) = T_a L(0) = T_a(0) = a$$

because the translation vector a must be the image of the zero vector by the affine map; thus $L = T_{-a}A$.

The *composite* of two affine maps $A = T_a L$ and $B = T_b M$ is also an affine map:

$$\begin{aligned} BA(x) &= T_b M T_a L(x) = b + M(T_a L(x)) \\ &= b + M(a + L(x)) \\ &= (b + M(a)) + ML(x) \end{aligned}$$

The composite BA thus consists of the linear map ML and the translation with the vector $b + M(a) = B(a)$.

EXAMPLE 6.3

Express the affine map $A = LT_b$ in the form $A = T_a L$ (with the same linear map L). What is the vector a?

6.2.1 Simple affine maps

The simplest affine map is, of course, the *translation*, as stated in equation (6.5). When acting upon an image, all its points are shifted by the vector $-a = (-a_1, -a_2)$.

For a *(point) symmetry* at the point (vector) $a = (a_1, a_2)$, the map is

$$x' = 2a - x \tag{6.7}$$

which is characterized by a being the mid-point between the vectors x and x'. The translation component is thus the vector $2a$, and the linear component consists of the point symmetry at $(0, 0)$, where $x' = -x$.

For a *line symmetry* about a line l, we assume l to be given by one of its points a and the unit vector e fixing its direction. The line is thus the set of points in the parametrization $x = a + te$. The vector a may be chosen freely among the points on the line, but one of these vectors will be perpendicular to l, that is, $a \cdot e = 0$. With this choice of a, the symmetry becomes

$$x' = 2a + 2(e \cdot x)e - x \tag{6.8}$$

It is characterized by the projection $a + (e \cdot x)e$ of x onto l being the mid-point between x and x'.

EXAMPLE 6.4
A line l is given by the point $a = (1, 0)$ and the direction vector $e = (\sqrt{2}/2)(1, 1)$. Find the vector a_0 on the line for which $a_0 \cdot e = 0$, and find the mirror image of the vector $(2, 0)$ in l as well. What are the six a-coefficients (cf. equation (6.6)) for the line symmetry in l?

A *rotation* D_θ through the angle θ (anticlockwise) about the origin is given by

$$x_1' = x_1 \cos\theta - x_2 \sin\theta, \qquad x_2' = x_1 \sin\theta + x_2 \cos\theta \tag{6.9}$$

(cf. Example 6.1). Using the shorthand $x' = D_\theta(x)$ for this linear map, the associated affine rotation around the point a is

$$x' - a = D_\theta(x - a) \tag{6.10}$$

With the same shorthand, this affine map may be written as $T_a D_\theta T_{-a}$.

Finally, another important affine map is the *multiplication* at point a. Here,

$$x' - a = k(x - a) \tag{6.11}$$

This map enlarges (or shrinks) vectors, radiating outwards from a, and the number k is called its *factor* or *magnification*.

Of the four simple linear maps described here, point and line symmetries, rotation, and multiplication, two are 'dispensible' in the sense that they may be constructed in terms of the others. A point symmetry may, for example, be considered as a rotation through 180° or as a multiplication with factor -1. More interestingly, any rotation may be decomposed into two line symmetries, the lines of which form half the rotation angle with each other. Properties like this are essential in the classification of these maps.

6.3 Similitudes

Among the line-preserving geometrical operations, an extremely important subclass exist which are called *similitudes* (Figure 6.2). These are characterized by conserving *distance ratios* as well. This property implies that if A, B, and C – arbitrary points in the original image – are mapped onto the respective points A', B', and C', then

$$\frac{|A'C'|}{|A'B'|} = \frac{|AC|}{|AB|} \tag{6.12}$$

Here, the modulus bars denote distance in the plane, that is, the two-dimensional norm.

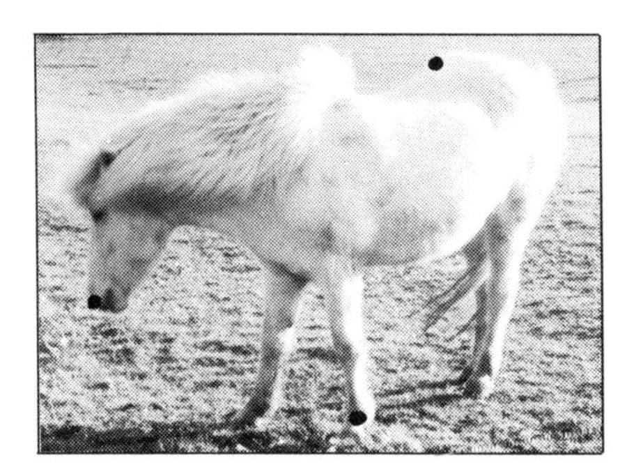

Figure 6.2 Similitude.

If equation (6.12) is expressed as $k = |A'B'|/|AB| = |A'C'|/|AC|$, the positive number k thus defined measures the ratio between the distances of two points A and B, before and after mapping, and this number is independent of the pair of points chosen. The number k is thus specific to the similitude, and this property may equally be expressed as follows:

$$|x' - y'| = k|x - y| \tag{6.13}$$

for all points (vectors) x and y, mapped onto x' and y'.

It should be noted that an arbitrary map, for which equation (6.13) holds, is automatically an affine map and consequently line-preserving. The latter attribute is thus a superfluous assumption. The reader is urged to investigate this point further....

From the relationship between norm and scalar product, that is,

$$|x - y|^2 = |x|^2 + |y|^2 - 2x \cdot y$$

scalar products are uniquely determined from the norm values. Consequently, for all x and y

$$x' \cdot y' = k^2 x \cdot y$$

Thus, not only are squared norms multiplied by k^2 in the similitude, but so also are scalar products.

Clearly, if two similitudes are composed, another will result, and the distance ratio of this similitude will be the product of the two original ratios. Also, the inverse affine map associated with a similitude is a similitude, and its ratio is the reciprocal of the original ratio.

In the following, we aim to provide a complete characterization of the similitudes. First, we examine linear similitudes, that is, affine maps without a translational component. For a vector x and its image x', this means

$$|x'| = k|x|$$

If $k = 1$, the similitude is *distance preserving*; it is also called a (linear) *isometry*. Obviously, if U is an isometry, then so is U^{-1}, and if V is another isometry, the composite VU of the two isometries U and V is also an isometry.

If, on the other hand, $k \neq 1$, the similitude may be decomposed into an isometry and a multiplication factor k (the order of composition is irrelevant): If A is a similitude of ratio k, then $(1/k)A$ is an isometry U, and

$$A = kU$$

This decomposition, too, is unique! That is, if $A = lV$ is another similitude of this type, then multiplication by a factor k/l must be an isometry VU^{-1}, whence $k = l$ and consequently $V = U$.

This characterizes the similitudes of ratio $k \neq 1$ via the similitudes of ratio 1, the isometries. The latter are, however, not difficult to describe. The key is the number of *fixed vectors* – vectors which are not altered by the map (and not equal to zero, as this vector is always unchanged under a linear map). We set out by investigating an isometry with:

One fixed vector. If the vector e is fixed, so are all vectors te, $t \in (-\infty, \infty)$, so we have an entire line consisting of fixed vectors. However, since distances are preserved, a vector outside the line is mapped either onto itself or onto its mirror image about the line. It will be seen from this that the map is either the identity (the trivial map having no effect) or the symmetry around the line defined by the fixed vector.

No fixed vectors. Here, we consider two vectors x and y which are transformed to x' and y', respectively. The angle between x and y (x' and y') is denoted θ (θ'). Since both norm and scalar product are conserved,

$$\cos\theta = \frac{x \cdot y}{|x||y|} = \frac{x' \cdot y'}{|x'||y'|} = \cos\theta'$$

We must have $\theta = \theta'$ (why?), and the angle between x and x' is equal to the angle between y and y'. Since x and y are arbitrary, the transformation must be a *rotation* through this angle.

This characterizes the linear isometries completely and, in consequence, the linear similitudes as well:

A linear isometry is either a line symmetry or a rotation.

A linear similitude is the composite of a multiplication and a linear isometry.

We are now in a position to return to the affine similitudes.

An affine similitude A with ratio $k \neq 1$ has exactly one fixed vector. This may be seen as follows. If $A = T_b L$ consists of the linear map L followed by the translation with vector b, then the fixed vector a must satisfy $L(a) - a = b$. But the only vector a solving the equation $L(a) - a = 0$ is the zero vector: from $a = L(a)$

it follows that $|a| = |L(a)| = k|a|$, hence $a = 0$. The equation $L(a) - a = b$ thus has a unique solution a (cf. Example 6.2). This point is called the *centre* of the similitude A, which is thus characterized as follows:

$$A(x) - a = L(x - a)$$

So, the translation component of A is defined by the vector $b = a - L(a)$, and the linear part L is the composite of a multiplication and a linear isometry.

Finally, the affine isometries are characterized as follows:

An affine isometry A, which is not a translation, is one of the following types:

1. $A(x) = D(x - a)$
   where a is a unique fixed point and D a (linear) rotation.
2. $A = T_b S_l = S_l T_b$
   where S is a symmetry about a unique line l, and b is a unique directional vector for this line.

The reader is urged to prove these results.

EXAMPLE 6.5

Three points $A = (0, 0)$, $B = (4, 0)$, and $C = (0, 3)$ are mapped by an affine similitude onto the points A', B', and C'. Describe the similitude in question in the following cases:

(a) $A' = (10, 2)$, $B' = (6, 2)$, $C' = (10, 5)$

(b) $A' = (3.6, 7.8)$, $B' = (10, 3)$, $C' = (0, 3)$.

6.4 Projective maps

A commonly occurring image-processing task is to 'raise' an image which has been photographed from an unfavourable position. For instance, a weather satellite might be located far away from the point where the action is, as in Figure 6.3. One might be interested in a reconstruction of the hypothetical image taken directly overhead. Even if the relevant information is present, in principle, in the 'tilted' image, the upright image will be of considerable interest for reference purposes. If it is to be stored for later comparison with other images of the same area, it might prove essential that the viewpoints are identical – for instance, a position at a given height vertically above the area.

Figure 6.4 shows how this problem (geometrical raising) is solved in practice. Problems of this type are generally solved from a detailed knowledge of the spatial geometry – angles, distances, and so on – describing the actual situation and the

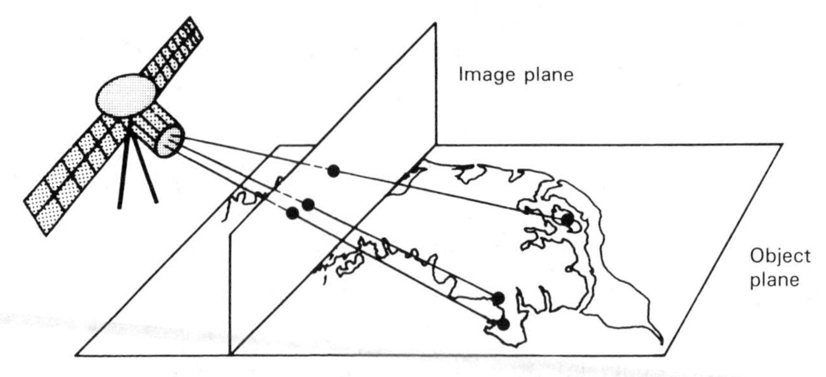

Figure 6.3 Satellite photography.

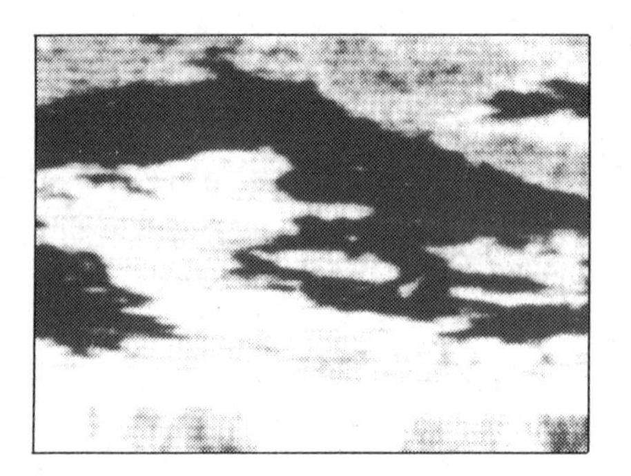
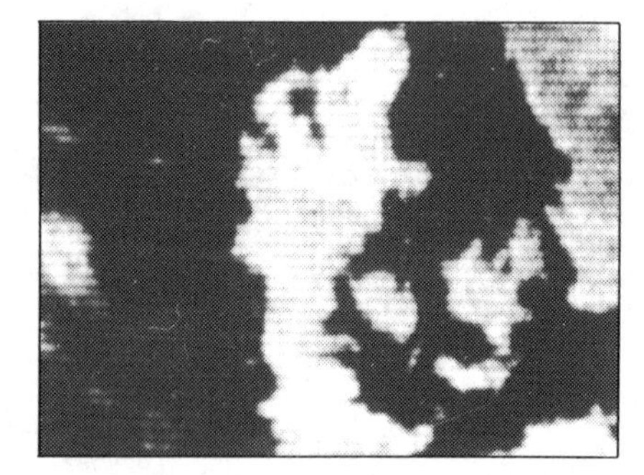

Figure 6.4 Geometric raising of image.

imaginary one. Then the actual image is transformed, pixel by pixel, into the upright one.

However, a few general principles will be of use for this reconstruction. As will be seen from Figure 6.5, the task is to perform a *central projection*. In a central projection, from the point O, of the plane p onto the plane p', the points $A, B, \ldots$ in p are mapped onto the intersection points $A', B', \ldots$ between $OA, OB, \ldots$ and p'. It is assumed that O is neither in p nor in p'. As was the case with the affine transformations, central projections are *line-preserving*. A line in p is contained in (precisely) one plane passing through O as well, and this plane will intersect p' in a straight line l' which consists of the images of the points in l.

There are two interesting exceptions. If l is located so as to make the plane through O and l *parallel* to p', l' is undefined. Conversely, it is impossible to have l' as an image line if the plane through O and l' is parallel to p.

However, it is possible to eliminate these exceptions if the planes p and p' are 'closed' by means of 'improper lines' to be envisaged as circles c and c' which, at

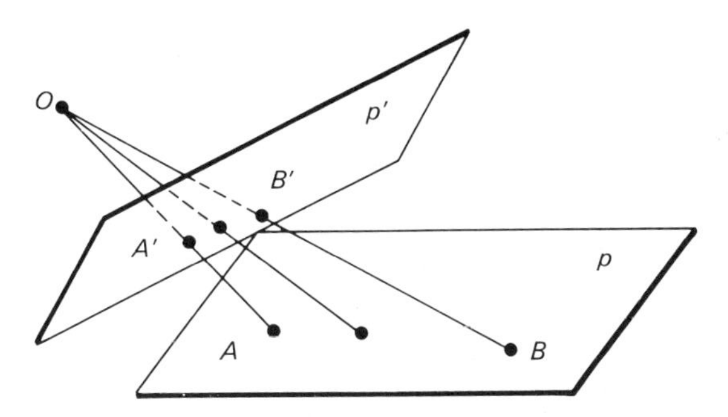

Figure 6.5 Central projection.

a very large distance, encircle the two planes. This produces a bijective correspondence between the lines mapped by the central projection. It is useful to think of the improper line in one plane appearing as a *horizon* in the other plane.

If the planes p and p' are parallel, the central projection of p onto p' is a similitude. This is the only type of projective map which preserves distance ratios. In the general case, however, another quantity is always conserved: the *cross ratio*. This is demonstrated in Figure 6.6, in which the plane of the page contains the projection point O as well as a line l and its image line l'.

The cross ratio $AB//CD$ for the four points A, B, C, and D on l is defined by

$$AB//CD = \frac{AC/AD}{BC/BD} = \frac{AC \cdot BD}{AD \cdot BC} \tag{6.14}$$

The line segments should be considered as signed according to a fixed orientation of l, even if the cross ratio is independent of this orientation. The following fundamental equality holds:

$$A'B//C'D' = AB//CD \tag{6.15}$$

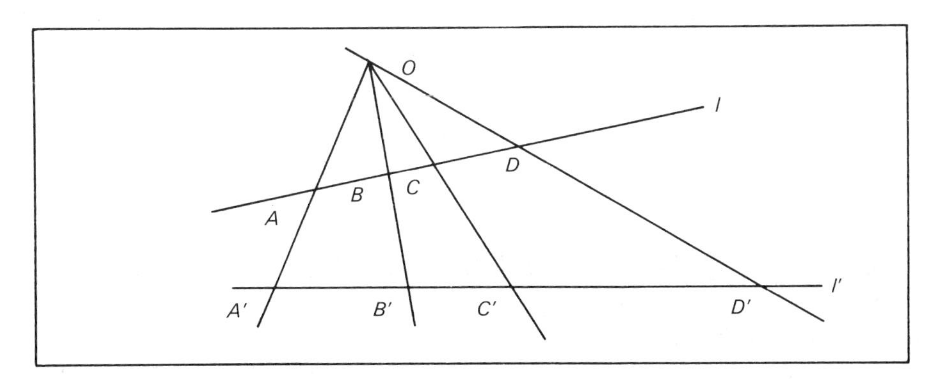

Figure 6.6 Invariance of the cross ratio.

This is seen by applying the sine relation from trigonometry to the triangles formed by O and the segments in question.

Central projections thus conserve both lines and cross ratios. Mappings of p onto p' possessing both of these properties are known as *collinear maps*, and central projections are thus collinear. The following main result is of great practical significance:

> A collinear map is uniquely determined by the specification of the images of four points (among which any three are not on a line).

This result is one of the cornerstones in the mathematical discipline called projective geometry. We shall not pursue this topic further; instead, this section concludes with a practical application.

EXAMPLE 6.6

A meteorologist studies a satellite image of Europe. In this image, the four cities Amsterdam, Berlin, Copenhagen and Dresden are identified, and their image coordinates are read off as listed in the table below.

City	Europe map	Image
Amsterdam	(170, 38)	(162, 15)
Berlin	(187, 623)	(−56, −11)
Copenhagen	(537, 546)	(64, −50)
Dresden	(20, 658)	(−98, 3)
?	?	(−132, 9)

In the table, the geographical positions of the cities are listed as well. To this end, a coordinate system centred upon Brussels is employed; listings are in kilometres.

A certain geographical feature which is not identifiable due to clouds, has image coordinates (−132, 9). Find the geographic position. (Using an atlas, identify the feature.)

Hint: Join pairs of points with straight lines, estimate cross ratios and use the fact that lines as well as cross ratios are identical in the two cases considered.

6.5 Interpolation

In order to deal with the various geometric operations as described, we shall need to *interpolate* in a digital image, that is, to convert it into an analog image. In Section 2.3 we mentioned a very rudimentary procedure, referred to as 'analog reconstruction', where the intensity was considered constant within each pixel. This conversion is of very limited use. First, the constructed image is discontinuous and thus physically unreasonable. Next, it does not meet the demand that the resolution in an interpolated image be increased – with reference to the general remark that if the sampling of the original image is satisfactory, then a slightly finer sampling will be in order as well (cf. Section 2.2). Nevertheless, the result of an interpolation will typically be a new digital image with unaltered resolution, as the new pixel grid has merely been shifted with respect to the first grid.

It has likewise already been mentioned (cf. Figure 2.11) that the 'analogization' is not unique. Even if the conversion is 'optimal', this notion will depend on the criteria employed (compare Examples 6.6 and 6.7 below).

In this section, we discuss three procedures, each of which has its pros and cons, but where the mathematical ingredients present ought to suffice for the development of similar methods designed for problems where certain quality criteria carry more weight than others. To begin with, we describe the interpolation problem in general.

6.5.1 Principles of interpolation

In an interpolation problem, one is asked to evaluate an analog signal $f(t)$ such that this signal assumes certain prescribed values $a_1, a_2, \ldots, a_N$ for $t = t_1, t_2, \ldots, t_N$. In the solution, one often limits the possible analog signals to a definite class of signal, for example polynomials. From these we select one assuming the value 0 at all sampling points t_n except one (say, t_m), where the value should be 1. If this signal is denoted $f_m(t)$, we have $f_m(t_n) = \delta[n - m]$. If such a signal is then selected for each $m \in [1, N]$, and the linear combination

$$f(t) = \sum_m a_m f_m(t)$$

is formed, then this signal f meets the requirements.

EXAMPLE 6.7 (Lagrange[1] interpolation)
The polynomial $P(t) = (t - t_1)(t - t_2) \ldots (t - t_N)$ is 0 at all points t_n. If the factor $(t - t_m)$ is removed, and if the remainder is called $P_m(t) = P(t)/(t - t_m)$, we still have $P_m(t_n) = 0$ for all n, except for $n = m$, where the value is

$$P_m(t_m) = (t_m - t_1) \ldots (t_m - t_{m-1})(t_m - t_{m+1}) \ldots (t_m - t_N)$$

Thus, the polynomial $P_m(t)/P_m(t_m)$ is of the above type $f_m(t)$, and the solution to the interpolation problem is

$$f(t) = \sum_m \frac{a_m}{P_m(t_m)} P_m(t) \tag{6.16}$$

[1] Joseph-Louis Lagrange, French mathematician, 1736–1813

Using this method, construct a third-order polynomial assuming the values listed below:

t_n	-1	0	1	2
a_n	2	0	-1	4

From this interpolation, find the value $f(\frac{1}{2})$.

Instead of polynomials, one quite often uses (especially for equidistant sampling) the signal sinc(t) (see p. 83) which is zero for all positive and negative multiples of π; for $t=0$, the value is unity. This means that if (for simplicity) $t_n = n\pi$, the above interpolation problem is solved by

$$f(t) = \sum_m a_m \frac{\sin(t - m\pi)}{t - m\pi} \tag{6.17}$$

EXAMPLE 6.8

Use sinc interpolation to determine an analog signal $f(t)$ having the following values:

t_n	$-\pi$	0	π	2π
a_n	2	0	-1	4

Evaluate $f(\frac{1}{2}\pi)$.

The latter interpolation method may, we note in passing, be constructed from Example 2.33 by Fourier transformation and use of Example 3.13. Finally, we note that an interpolation method was encountered in connection with Example 2.40 and the preceding remarks, as well as in Example 2.41.

6.5.2 Bilinear smoothing

This refers to smoothing of the reconstructed analog image, and we consider the 2×2 image section shown in Figure 6.7. This is perceived as a reconstructed analog image, that is, an analog image of constant intensities within each of the four pixel areas. We now evaluate the integrated intensity in the shaded pixel, which is centred on (x, y) (where $|x|, |y| \leqslant \frac{1}{2}$). The method amounts to determining the overlap

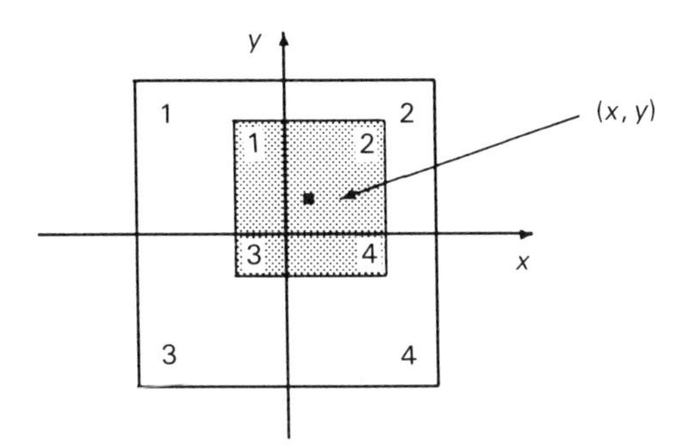

Figure 6.7 Interpolation in a 2×2 image area.

of this pixel with the four given ones and summing the corresponding intensity contributions proportionally. The overlaps are given below:

Pixel no.	Intensity	Overlap
1	n_1	$\frac{1}{4}-\frac{1}{2}x+\frac{1}{2}y-xy$
2	n_2	$\frac{1}{4}+\frac{1}{2}x+\frac{1}{2}y+xy$
3	n_3	$\frac{1}{4}-\frac{1}{2}x-\frac{1}{2}y+xy$
4	n_4	$\frac{1}{4}+\frac{1}{2}x-\frac{1}{2}y-xy$

On pairwise multiplication of the items in the last two columns and summing of the products, we arrive at the interpolation formula sought:

$$b_{xy} = A + Bx + Cy + Dxy \tag{6.18}$$

the coefficients A, B, C, and D of which are as follows:

$$\begin{aligned} 4A &= +n_1+n_2+n_3+n_4 \\ 2B &= -n_1+n_2-n_3+n_4 \\ 2C &= +n_1+n_2-n_3-n_4 \\ D &= -n_1+n_2+n_3-n_4 \end{aligned} \tag{6.19}$$

Here, the total intensity in the new pixel is approximated by the simple bilinear polynomial in equation (6.18). This polynomial may be regarded as the result of a *convolution* of the reconstructed analog image and the rectangular image $r_{1,1}(x, y)$ (cf. Section 2.4). One might also say that b_{xy} is the result of a *resampling*, using averages, of the reconstructed analog image. Note that equation (6.18) may be regarded both as one pixel value in a digital image and as an analog image, and that the pointwise sampling of the latter is identical to that obtained by averaging; both methods yield the digital (resampled) value.

EXAMPLE 6.9

In a digital image, one encounters the following 2×2 pixels:

37	43
41	39

In the interpolation method described, $b_{xy} = 40 + 2x + 8xy$. The interpolated intensity in a new (centred) pixel (with $x = y = 0$) thus equals the average, 40, of the four pixel values.

6.5.3 Linear maximum likelihood interpolation

In this method, we interpolate over 2×2 image sections, as in Figure 6.7. For now, however, we assume the intensity over the section to be given by the analog image $\bar{b}(x, y)$, $x \in (-1, 1)$, $y \in (-1, 1)$. The number of photons in a 1×1 area centred on (x, y) is thus

$$b(x, y) = \int_{x-1/2}^{x+1/2} \int_{y-1/2}^{y+1/2} \bar{b}(x, y)\, \mathrm{d}x\, \mathrm{d}y$$

The four photon counts should, of course, be regarded as mean values for four Poisson distributions, for which four observations n_1, n_2, n_3, and n_4 are available.

The method is based upon the assumption that $\bar{b}(x, y)$ is a first-order polynomial:

$$\bar{b}(x, y) = A + 2Bx + 2Cy$$

so that the mean values for these four pixels are

$$\begin{array}{ll} b_1 = A - B + C & b_2 = A + B + C \\ b_3 = A - B - C & b_4 = A + B - C \end{array}$$

The likelihood function, that is, the probability of having observed precisely n_1, n_2, n_3, and n_4, thus becomes

$$\mathscr{L} = \mathrm{e}^{-b_1} \frac{b_1^{n_1}}{n_1!} \mathrm{e}^{-b_2} \frac{b_2^{n_2}}{n_2!} \mathrm{e}^{-b_3} \frac{b_3^{n_3}}{n_3!} \mathrm{e}^{-b_4} \frac{b_4^{n_4}}{n_4!}$$

from which

$$L = \ln \mathscr{L} = -\sum_i b_i + \sum_i n_i \ln b_i - \sum_i \ln(n_i!)$$

Here, $\Sigma\, b_i = 4A$, so that the ML estimators for A, B, and C are found from the equations

$$\frac{\partial L}{\partial A} = +n_1/b_1 + n_2/b_2 + n_3/b_3 + n_4/b_4 - 4 = 0$$

$$\frac{\partial L}{\partial B} = -n_1/b_1 + n_2/b_2 - n_3/b_3 + n_4/b_4 = 0$$

$$\frac{\partial L}{\partial C} = +n_1/b_1 + n_2/b_2 - n_3/b_3 - n_4/b_4 = 0$$

From the last two equations it follows that

$$\frac{n_1}{b_1} = \frac{n_4}{b_4} \qquad \text{and} \qquad \frac{n_2}{b_2} = \frac{n_3}{b_3}$$

This further leads to

$$\left.\begin{matrix} B \\ C \end{matrix}\right\} = \frac{1}{2}\left(\frac{n_2 - n_3}{n_2 + n_3} \pm \frac{n_4 - n_1}{n_4 + n_1}\right)A \tag{6.20}$$

and finally

$$A = \tfrac{1}{4}(n_1 + n_2 + n_3 + n_4) \tag{6.21}$$

EXAMPLE 6.10

In pixels 1–4 of Figure 6.7, 60, 31, 38, and 41 counts, respectively, are recorded. What is the value for a pixel centred on (0.25, 0.25), if the above interpolation method is employed?

Since the above method takes into account the exact Poisson distribution, it is especially suited for cases where the numbers of photons are small. For higher intensities and in other situations where the Poisson distribution may be approximated by the normal distribution, a method like that in the next and concluding subsection may be applied.

6.5.4 Interpolation by signal distance

In Figure 6.8 we interpolate over the nine nearest neighbour pixels. As before, we assume that the 'correct' intensity distribution across the area is an analog image $\bar{b}(x, y)$, $x \in (-1.5, 1.5)$, $y \in (-1.5, 1.5)$, $\bar{b}$ being a second-order polynomial:

$$\bar{b}(x, y) = A_1 + A_2 x + A_3 y + A_4(3x^2 - \tfrac{9}{4}) + A_5(3x^2 - \tfrac{9}{4}) + A_6 xy \tag{6.22}$$

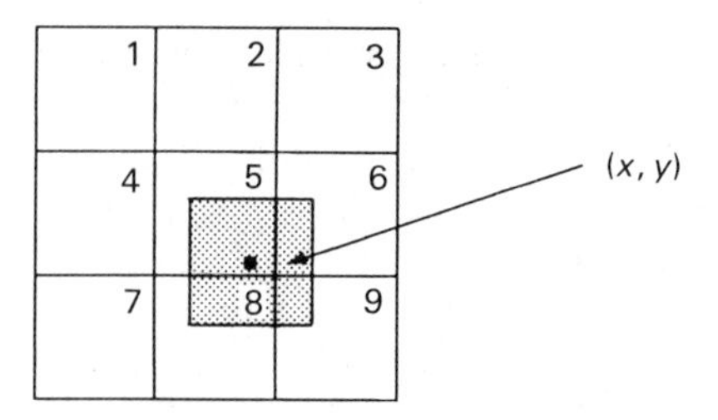

Figure 6.8 Interpolation area.

(The significance of the rather curious choice of the coefficients A_4 and A_5 – and A_1 – will become clear in a moment.)

With this intensity distribution across the 3×3 image, a pixel centred on (x, y) would record a total intensity of

$$b(x, y) = \int_{y-1/2}^{y+1/2} \int_{x-1/2}^{x+1/2} \bar{b}(x, y) \, \mathrm{d}x \, \mathrm{d}y$$

$$= A_1 + A_2 x + A_3 y + A_4(3x^2 - 2) + A_5(3y^2 - 2) + A_6 xy$$

On substitution of $(x_1, y_1) = (-1, 1)$, $(x_2, y_2) = (0, 1)$, ..., $(x_9, y_9) = (1, -1)$, nine 'ideal' values, $b_1, \ldots, b_9$, are obtained. The bs will, of course, depend on the A-coefficients. In the nine pixels, however, the counts $n_1, \ldots, n_9$ have been observed. Our task is to determine the six coefficients $A_1, \ldots, A_6$, for which the signal distance $|n - b|$ given by

$$|n - b|^2 = \sum_{k=1}^{9} (n_k - b_k)^2$$

is a minimum.

If we return to the issue of signal approximation from Section 2.6, we find once again that the solution consists in a projection: If we construct the following six vectors, having nine components each,

$$f_1 = (1, 1, \ldots, 1)$$
$$f_2 = (x_1, \ldots, x_9)$$
$$f_3 = (y_1, \ldots, y_9)$$
$$f_4 = (3x_1^2 - 2, \ldots, 3x_9^2 - 2)$$
$$f_5 = (3y_1^2 - 2, \ldots, 3y_9^2 - 2)$$
$$f_6 = (x_1 y_1, \ldots, x_9 y_9)$$

we must determine that linear combination $b = \Sigma_{k=1}^{6} A_k f_k$, for which the distance from $n = (n_1, \ldots, n_9)$ is a minimum. If the f-vectors are written in full:

$$
\begin{array}{lrrrrrrrrr}
f_1 = (& 1 & 1 & 1 & 1 & 1 & 1 & 1 & 1 & 1) \\
f_2 = (& -1 & 0 & 1 & -1 & 0 & 1 & -1 & 0 & 1) \\
f_3 = (& 1 & 1 & 1 & 0 & 0 & 0 & -1 & -1 & -1) \\
f_4 = (& 1 & -2 & 1 & 1 & -2 & 1 & 1 & -2 & 1) \\
f_5 = (& 1 & 1 & 1 & -2 & -2 & -2 & 1 & 1 & 1) \\
f_6 = (& -1 & 0 & 1 & 0 & 0 & 0 & 1 & 0 & -1)
\end{array}
$$

they are easily seen to be orthogonal. According to equation (2.28), we then have

$$A_k = \frac{n_k \cdot f_k}{|f_k|^2}$$

These coefficients are collected into the following array:

$$
\begin{array}{rlllllllll}
9A_1 = & +n_1 & +n_2 & +n_3 & +n_4 & +n_5 & +n_6 & +n_7 & +n_8 & +n_9 \\
6A_2 = & -n_1 & & +n_3 & -n_4 & & +n_6 & -n_7 & & +n_9 \\
6A_3 = & +n_1 & +n_2 & +n_3 & & & & -n_7 & -n_8 & -n_9 \\
18A_4 = & +n_1 & -2n_2 & +n_3 & +n_4 & -2n_5 & +n_6 & +n_7 & -2n_8 & +n_9 \\
18A_5 = & +n_1 & +n_2 & +n_3 & -2n_4 & -2n_5 & -2n_6 & +n_7 & +n_8 & +n_9 \\
4A_6 = & -n_1 & & +n_3 & & & & +n_7 & & -n_9
\end{array}
$$

which, together with equation (6.22), is the solution to the interpolation problem.

EXAMPLE 6.11

A 3×3 digital image consists of the numbers

40	19	33
42	28	40
38	30	35

The coordinate system is placed at the centre of the image. Approximate the image with a second-order polynomial $\bar{b}(x, y)$ as described. With this, find the (point) intensity at $(\frac{1}{2}, \frac{1}{2})$, as well as the total intensity of a pixel centred here – that is, find the quantities $\bar{b}(\frac{1}{2}, \frac{1}{2})$ and $b(\frac{1}{2}, \frac{1}{2})$.

7

Image Improvement

The most interesting aspect of image processing is the existence of a rapidly expanding set of techniques, collectively referred to as 'image improvement'. In the application of these techniques, the raw material is assumed to be an image, or images, in which an interesting amount of information is present; this information, however, is presented in a way unsuitable for the human eye. We aim to develop methods for extracting this information, that is, for 'improving' the images. Three examples follow, one for each section in the present chapter.

Figure 7.1a is an image with poor contrast. If one uses a smaller dynamic range and manipulates the grey level scale within it, Figure 7.1b results – an example of a simple operation which might be called 'level-dependent quantization'. The traditional term is *image enhancement*. More subtle methods will be presented shortly.

Figure 7.2a is superimposed on a periodic pattern. After Fourier transformation of the image and removal of the components corresponding to this pattern, the

(*a*)

(*b*)

Figure 7.1 Image enhancement.

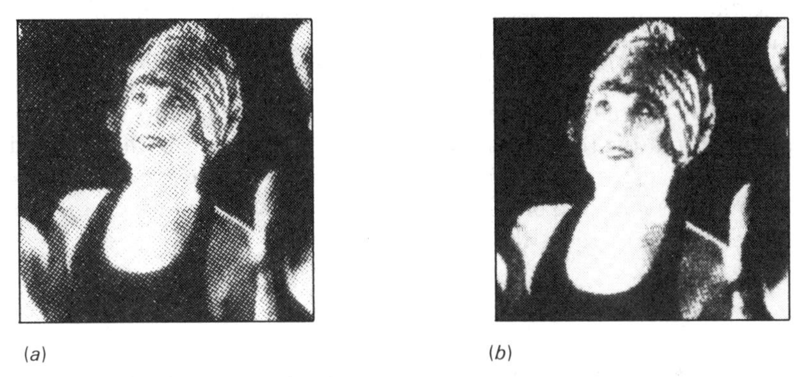

(a) (b)

Figure 7.2 Filtering.

(a) (b)

Figure 7.3 Image restoration.

Figure 7.4 Eyesight test.

image resulting from an inverse Fourier transformation is Figure 7.2b. The undesirable contribution to the image is thus *filtered* out.

Figure 7.3a (see previous page) is a photograph of a car speeding away from a bank robbery. The photographer, however, did not manage to set the exposure time properly. But using the methods of Section 7.3, Figure 7.3b was obtained from the blurred image, after which the perpetrator was identified.

These methods may also be applied to Figure 7.4 (see previous page), where the photographer forgot to focus the camera.

7.1 Level operations and image contrast

7.1.1 Simple level operations

The simplest level operations are those for which the change in pixel values does not depend upon position within the image. The quantity $b(x, y)$ is thus changed to $b'(x, y)$, where the change depends only upon the level itself:

$$b' = f(b) \tag{7.1}$$

Normally, we require the function f to be increasing, that is,

$$f(b_1) \leqslant f(b_2) \qquad \text{if} \qquad b_1 < b_2$$

since, otherwise, the relative intensities in the image will be interchanged. The function might possibly be decreasing, with a negative image as the result.

Among these level operations, the simplest are the (piecewise) linear ones, where the function f is continuous and of the form

$$f(b) = A + Bb \tag{7.2}$$

with different pairs of constants A and B in different level intervals. One example is the level operation to be performed when an image is displayed on a computer screen. Here, a piecewise linear function is applied (see Figure 7.5).

The most commonly applied principle is the following. If the dynamic range of the image is defined by $b_{\min}$ and $b_{\max}$, we choose two 'cut' levels b_- and b_+ such that $b_{\min} \leqslant b_- < b_+ \leqslant b_{\max}$, after which the levels in the interval from b_- to b_+ are distributed within the dynamic range for the screen. If this range consists of the grey levels from 1 to N, the '1' is usually reserved for levels between $b_{\min}$ and b_-, whereas N comprises the levels between b_+ and $b_{\max}$. This was done when going from Figure 7.1a to Figure 7.1b.

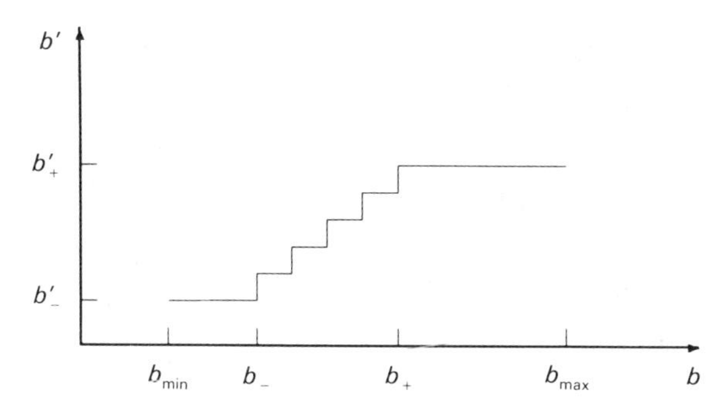

Figure 7.5 Piecewise linear function for image presentation.

Another frequently used level operation is the logarithm:

$$b' = \log b \tag{7.3}$$

or its inverse, the exponential:

$$b' = \exp b \tag{7.4}$$

where the bases for the log and exp functions are arbitrary.

The choice of these operations is often dictated by physical considerations. For instance, one may wish to take into account – or to correct for – the logarithmic sensitivity of the eye (Section 1.4). Also, certain types of image (such as X-ray images) in their original form *are* modified by a logarithmic or exponential law. The reason is that radiation, on passing through an absorbing medium, is generally reduced according to the formula

$$I = I_0 e^{-kz}$$

where the intensity I has the value I_0 at the point $z = 0$; on propagation in the z-direction, the ratio k is absorbed per unit length (cf. Section 7.5). The quantity kz is called the *optical path*. The logarithm of the intensity after passage (Figure 7.6) is thus an image measuring the physical thickness of, or the optical path through, the absorbing material.

Finally, one might wish to correct for the sensitivity of a photographic image. The blackening of a film or photographic plate – that is, the grain density – is only approximately proportional to the amount of light; the quantum efficiency (cf. p. 11) depends, in other words, upon intensity. Figure 7.7 shows a typical photographic sensitivity curve. Note that the horizontal axis gives the logarithm of intensity – even in the linear range in the figure, proportionality between blackening and intensity does not hold exactly. However, an exponential level transformation (equation (7.4)) in this range will result in a satisfactorily linear relation between level and physical light intensity.

(a) $b(x, y)$

(b) $\log b(x, y)$

Figure 7.6 Logarithmically corrected image.

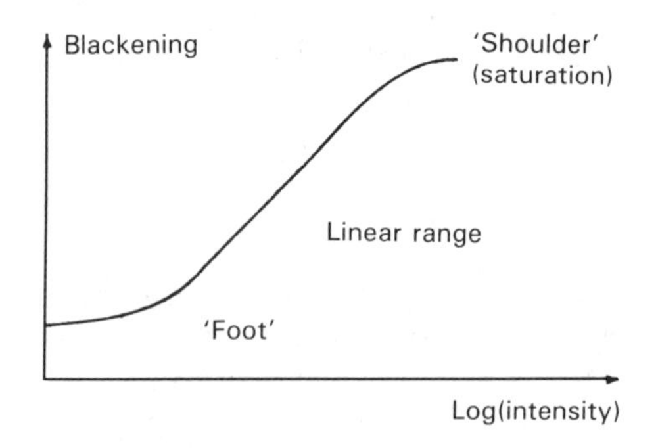

Figure 7.7 Photographic sensitivity curve (Hurter–Driffield curve).

In general, the conversion from image level b to physical intensity I is of the form

$$I = f^{-1}(b)$$

where f^{-1} is the inverse of the sensitivity function f.

Among the blessings of the CCD camera is a very exact proportionality between level and intensity over a large dynamic range. Thus, the only relevant sensitivity correction for a CCD chip will be

$$I = Ab \tag{7.5}$$

where the physical intensity I, for instance, is given as photon counts, and where b is the level measured in the readout units determined by the manufacture of the chip. The conversion factor A often varies slightly between the pixels, and if this variation is to be taken into account, equation (7.5) should be modified as follows:

$$I(x, y) = A(x, y)b(x, y) \tag{7.6}$$

If this function $A(x, y)$ is not specified by the manufacturer of the chip, the usual remedy is to obtain an exposure b_0 of a uniformly illuminated surface, where $I(x, y) = I_0$, so that

$$I_0 = A(x, y)b_0(x, y) \tag{7.7}$$

From this *flat field exposure*, the spatial sensitivity function $A(x, y)$ is determined (apart from a multiplicative constant); then, equations (7.6) and (7.7) yield

$$I(x, y) = I_0 \frac{b(x, y)}{b_0(x, y)} \tag{7.8}$$

In contrast to equation (7.1), this type of operation does not depend explicitly upon level but upon position. In general, a level operation is of the form

$$b'(x, y) = f(x, y, b(x, y))$$

thus, a combination of the two simple varieties. These types are referred to as, respectively, *global* and *local* level operations. For example, the global operation (7.1) becomes local if applied to a small image segment only.

7.1.2 Image contrast and histogram equalization

The *contrast* of an image is given by the distribution of its grey levels – or, for digital images, its pixel values. If this distribution is concentrated near a certain level, then the contrast is obviously low. If, on the other hand, a wide range of levels is represented, then the contrast is high.

The contrast of an image as a whole is given by its histogram, and contrast manipulations within the image should be reflected in corresponding changes in the histogram. If an image has been segmented according to level, these contrast operations may be applied to the corresponding parts of the histogram only.

If, for instance, the level distribution is concentrated around two or more levels, the contrast of the whole image may well be described as 'high'. But if the image is segmented between these levels, the contrast within the individual segments will be low, and the contrast of the image as a whole will be given by the variations between these segments.

From these qualitative considerations, it seems plausible that the uniform level distribution must have some special status in this context and describe a situation with a maximum contrast. There are obvious echoes here of the information–theoretical point of view adopted in Chapter 5. In Example 4.6 it was pointed out that among the various distributions on $[1, N]$, the uniform distribution was the one which maximized the information content or *entropy*. This suggests a very close relationship between image contrast and entropy of the level distribution.

In image processing, the task of maximizing contrast is known as *histogram equalization*. Here, we perform a global level operation, with a uniform level distribution as the intended effect.

If this level operation is called f, changing the level b into $b' = f(b)$, the original histogram h is changed into a new one, h'. Assuming, first, the level distribution to be continuous, let $h(b)$ and $h'(b')$ denote the corresponding probability densities. We also assume that the dynamic range both before and after the operation is the interval (0, 1); this may otherwise be achieved by combining f with linear level operations. The new histogram h' is thus required to be the constant function equal to unity in (0, 1).

According to Appendix A, the relation between $h(b)$ and $h'(b')$ is given by

$$h'(b')\ \mathrm{d}b' = \mathrm{d}b' = h(b)\ \mathrm{d}b$$

that is, since $\mathrm{d}b'/\mathrm{d}b = \mathrm{d}f/\mathrm{d}b$,

$$\frac{\mathrm{d}f}{\mathrm{d}b} = h(b)$$

Histogram equalization is thus achieved by choosing the level transformation function as the integral of the probability density, that is, as the (cumulative) distribution function

$$f(b) = \int_0^b h(x)\ \mathrm{d}x \tag{7.9}$$

EXAMPLE 7.1
An image has histogram $h(b) = 2\sqrt{b}$, $b \in (0, 1)$. The level transformation equalizing this histogram is given by $f(b) = 1/\sqrt{b}$.

In real life, histograms are not continuous. If a histogram is specified as a probability distribution p_n, $n \in [0, N-1]$, with $\Sigma_n\, p_n = 1$, we may nevertheless envisage the histogram values p_n to be defined at the points $b_n = n/N$, all within the interval (0, 1). In order to equalize this histogram, it seems reasonable, in place of equation (7.9), to employ

$$f(b) = \sum_{n \leqslant b} p_n, \qquad b \in (0, 1) \tag{7.10}$$

This method has been brought into play in Figure 7.8, which shows (a) the histogram (probability distribution) p_n, (b) the equalizing function f, and (c) the equalized histogram p_n'.

EXAMPLE 7.2
Carry out the histogram equalization as shown in Figure 7.8, where the p_n-values are, respectively, 0.15, 0.08, 0.20, 0.07, 0.35, and 0.15 (that is, find the values p_n').

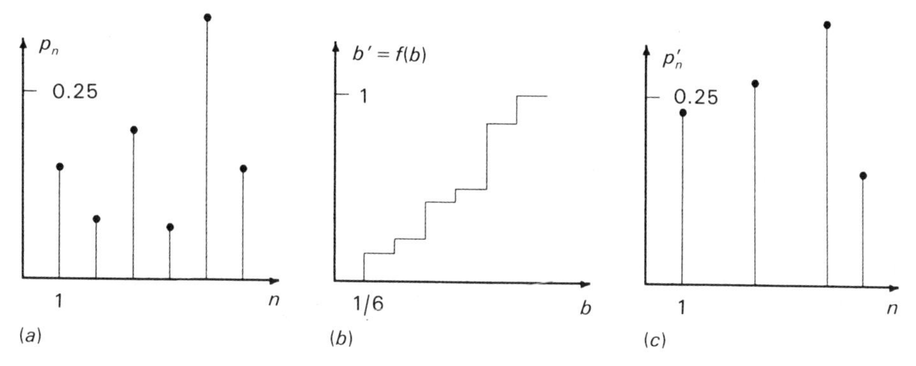

Figure 7.8 Histogram equalization.

EXAMPLE 7.3
An image has been quantized with eight levels $n = 0, 1, \ldots, 7$, the frequencies of which are $p_n = a\sqrt{n + 0.5}$. Find a. Equalize the histogram as described above. What is p'_4?

Clearly, a perfectly equalized histogram should never be expected. This might be ascribed to the discontinuity of the equalizing function f. Several remedies exist, but there is a tendency to prefer simple procedures and approximately equalized histograms to these more sophisticated methods, if the price to be paid is a drastic increase in computing time.

In some cases, one would want the image contrast to be given by some particular histogram, not necessarily that of the uniform distribution. This contrast manipulation is called *histogram specification*. If h_1 is the original histogram, to be changed into h_2, we use the notation

$$f_1(b) = \int_0^b h_1(x)\,\mathrm{d}x, \qquad f_2(b) = \int_0^b h_2(x)\,\mathrm{d}x$$

where, for simplicity, only functions on (0, 1) are considered. The change desired now follows from the level operation

$$f = f_2^{-1} \circ f_1$$

since f_1 changes h_1 into the uniform distribution, which again is changed into h_2 by f_2^{-1}, the inverse function of f_2.

EXAMPLE 7.4

Which transformation should be applied in order to change the histogram $h_1(b) = \frac{3}{2}\sqrt{b}$ into

(a) $h_2(b) = 3b^2$ (b) $h_2(b) = \frac{\pi}{2}\sin(\pi b)$?

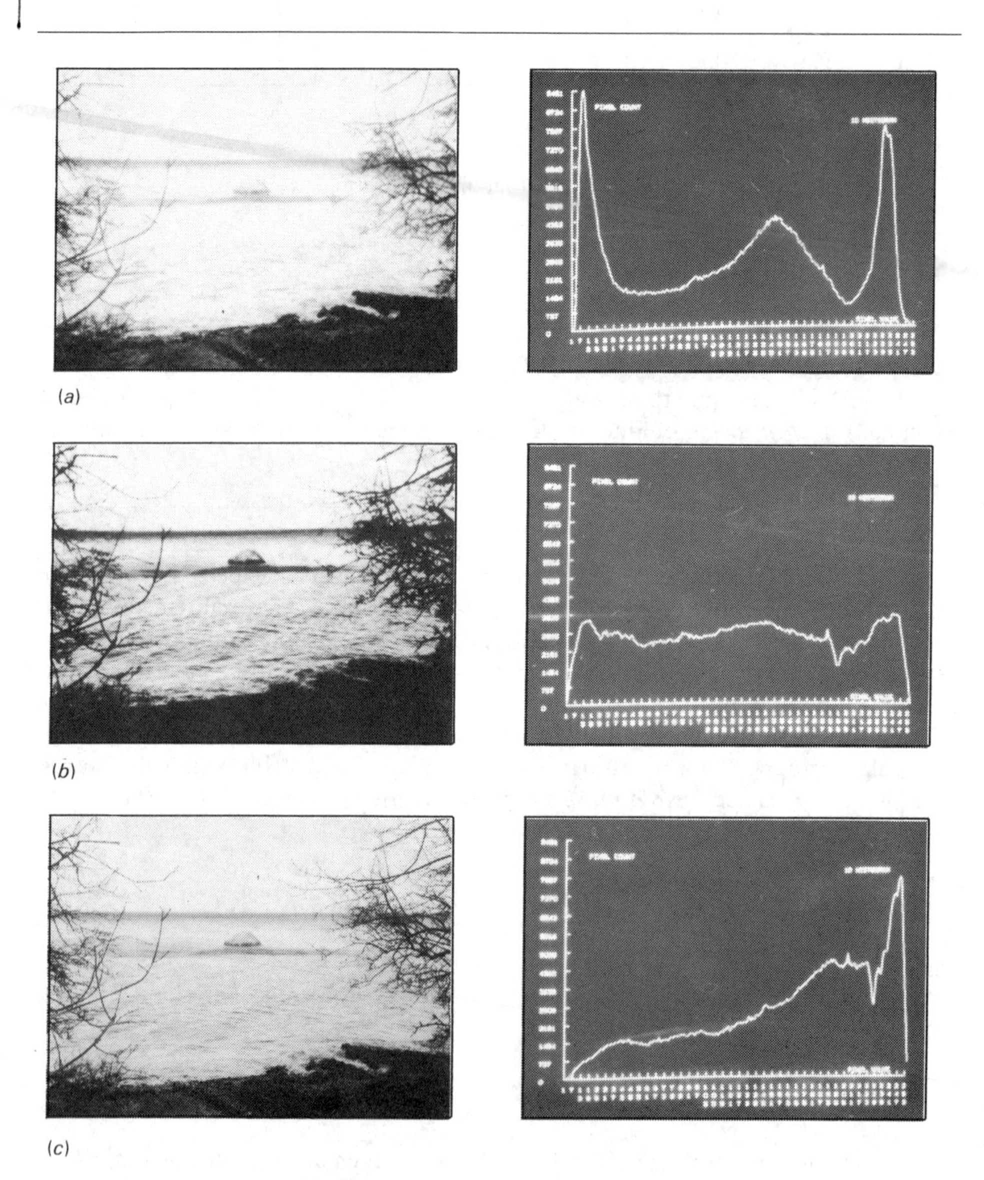

Figure 7.9 Contrast manipulations.

Figure 7.9 illustrates the effect of histogram equalization and histogram specification. Figure 7.9a shows an image and, to its right, its histogram. Figure 7.9b shows the effect of histogram equalization, while Figure 7.9c shows the specification of the histogram $h(b) = e^b/(e-1)$. It will be evident that, even if the histograms are imperfect, the changes in contrast are considerable.

7.1.3 Variance equalization

The method developed in Section 7.1.2 to change the contrast in an image to meet particular ends, resulted in global level operations $b' = f(b)$ having or approximating the effect desired. If this procedure is extended to one or more segments, separately, in an image, the scope is broadened considerably. The global operation $b' = f(b, x, y)$ now becomes a local, albeit slowly varying function of the image coordinates x and y.

In practice, this method often puts considerable strain on available computing resources. A compromise consists in the use of the *image variance* over the segment as a measure of its contrast; then, a uniform variance across the entire image is aimed at (*variance equalization*).

The application of this method is illustrated in Figure 7.10, which derives from a simple version of this technique. The linear operation $b' = A + Bb$ (equation (7.2)) was employed, with slightly position-dependent coefficients A and B obtained from the requirement that the new average $\bar{b}'$ over the segment in question be equal to the original one, $\bar{b}$, and that the new variance σ'^2 should equal a prescribed value σ_0^2. This is achieved by choosing

$$b' = \bar{b} + (b - \bar{b})\sigma_0/\sigma \tag{7.11}$$

(a)

(b)

Figure 7.10 Variance equalization.

It is often a matter of taste whether histogram equalization or variance equalization should be preferred for the solution of a specific problem concerning image (contrast) enhancement.

7.2 Filtering

7.2.1 Ideal filters

In the type of signal processing known as *filtering*, the aim is to remove certain (harmonic) components of a signal and to retain others. Two main types are known as, respectively, *lowpass* and *highpass filtering*, according to the frequencies of those components allowed through the filter. The term 'filter' is, in this context, roughly synonymous with the term 'system' as used in Section 3.4, but it refers in particular to situations where the system function vanishes in certain frequency ranges and is approximately constant for other frequencies.

The term is, moreover, preferable for systems with a certain correcting effect, where a specific undesired signal or image property is to be eliminated. An image like Figure 7.4, being out of focus, may perfectly well be referred to as being lowpass filtered. In this case, however, the filtering effect would normally be considered undesirable, and the image may be corrected by application of inverse filtering or image restoration (see Section 7.3). The traditional usage is manifestly unclear, but should hopefully not cause too much confusion.

Figure 7.11 shows the system function for three types of ideal one-dimensional filters. Besides lowpass and highpass filters, it also shows an ideal *bandpass filter*, the purpose of which is to allow the passage of signal components with frequencies in a given interval, while components having frequencies outside the interval are rejected entirely. The figure only shows the amplitude part $|H|$ of the system function H, since the phase part vanishes by definition. The filter action consists, as usual, in a multiplication of the spectrum F for the input signal f by the filter function H, producing the spectrum G for the output signal:

$$G(\omega) = H(\omega) = |H(\omega)|\,e^{i\phi}F(\omega)$$

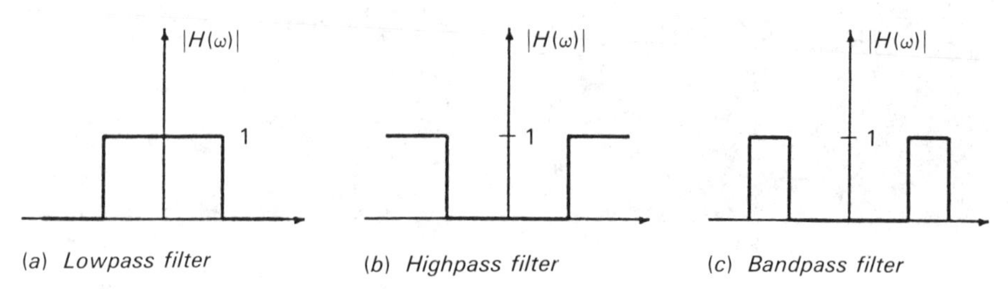

(a) *Lowpass filter* (b) *Highpass filter* (c) *Bandpass filter*

Figure 7.11 Ideal filters.

that is,

$$G(\omega) = |H(\omega)|\, F(\omega) \tag{7.12}$$

Thus, the filtered signal (image) is constructed from the original one according to the well-known recipe: Fourier transformation, multiplication, and inverse Fourier transformation.

Note that the system function is *even*, that is, symmetrical about the vertical axis $\omega = 0$, and that negative frequencies are present. This is due to the notation involving complex-valued signals, since a real harmonic signal of a given frequency ω is the sum of two complex harmonic signals, the frequencies of which are ω and $-\omega$.

Finally, it should be mentioned that a filter having a filter function of type $1 - b(\omega)$, $b(\omega)$ being the bandpass filter (Figure 7.11c), is called a *bandstop filter*.

Figure 7.12a shows the amplitude spectrum for the image shown in Figure 7.2a (the phase spectrum has not been shown). There is no difficulty in identifying the components due to the periodic pattern in the original. Next, a corresponding bandstop filter is constructed, blocking the frequencies in question (Figure 7.12b); the filter value 1 is denoted by white, the value 0 by black. The filtered amplitude spectrum, that is, the product of the images in Figures 7.12a and 7.12b, is shown in Figure 7.12c.

The improved image in Figure 7.2b now results from inverse Fourier transformation of the complex image whose amplitude part equals the product Figure 7.12c and whose phase spectrum is identical to the original one. Instead of the circular bandstop areas in Figure 7.12b, one may employ the computationally slightly more convenient rectangular bandstop.

Even if the filtering principle applies to many problems in image processing, the two main types mentioned at the beginning of this section are almost exclusively used for one specific purpose each: lowpass filtering is used for *smoothing*, in particular *elimination of image noise*; and highpass filtering is used for *contour enhancement*. This is because normal image information is represented in a regular, uniform fashion without large intensity variations between neighbouring pixels (cf.

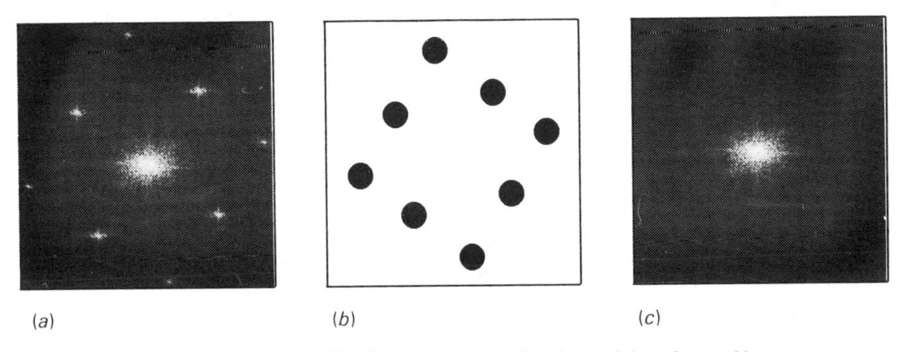

Figure 7.12 Amplitude spectrum with adapted bandstop filter.

pp. 3–5). In the frequency domain, this type of image information is represented by low (spatial) frequencies. Sharp transitions in the image – whether due to edges or image noise – will be represented by high-frequency components.

A lowpass filtered image is therefore smoothed; the noise will be reduced, but the contours will be blurred as well. In a highpass-filtered image, the contours will stand out clearly – as will the image noise. Many of these drawbacks may be diminished by appropriately chosen bandpass or bandstop filtering, and the choice of filter for a given purpose is an inevitable compromise between the image quality desired and the accompanying drawbacks.

Figure 7.13 illustrates the effect of the two types of filtering. The original image (Figure 7.13a) has been subjected to filtering corresponding to a certain cut-off frequency ω_c, so that Figure 7.13b displays the result of rejecting all frequencies above ω_c; in Figure 7.13c, all frequencies below another frequency ω_c have been rejected.

In Figure 7.13b we note a characteristic periodic pattern, called *ringing*. This is due to the sharp 'edges' of the filter employed; they disappear if a smoother system function is used. The filter shown in Figure 7.14 (named after Julius von Hann) is commonly used for smoothed lowpass filtering. Its effect is illustrated in Figure 7.15. Instead of a product of two one-dimensional filters, one most often uses a *circular* Hanning filter in image processing (this was, in fact, done in Figure 7.15). This filter is defined by

$$H(u, v) = \begin{cases} \frac{1}{2} + \frac{1}{2}\cos(\pi\omega/2\omega_c) & \text{for } \omega^2 = u^2 + v^2 < \omega_c^2 \\ 0 & \text{otherwise} \end{cases} \tag{7.13}$$

where ω_c is the characteristic radius of the Hanning filter, in the frequency domain.

Two other common filters are the *Butterworth filter*, that is,

$$H(u, v) = \frac{1}{1 + (\omega^2/\omega_c^2)^n} \tag{7.14}$$

(a) Original image

(b) Lowpass filtering

(c) Highpass filtering

Figure 7.13 Image filtering.

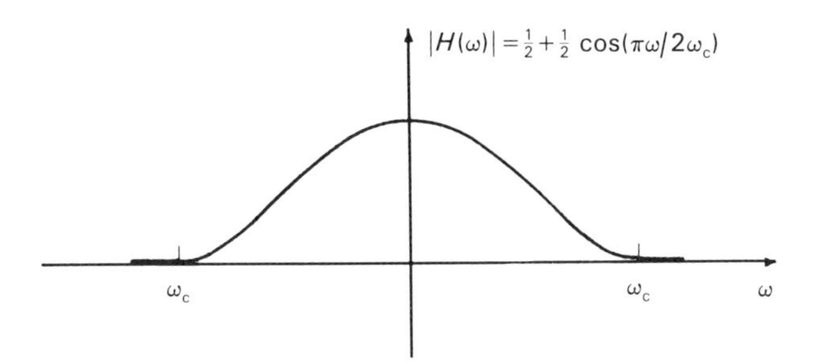

Figure 7.14 Hanning filter.

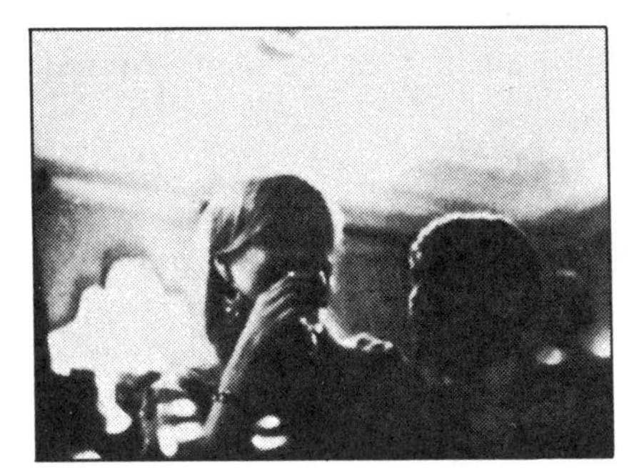

Figure 7.15 Hanning filtered image.

and the *exponential filter*

$$H(u, v) = e^{-(\omega^2/\omega_c^2)^n} \tag{7.15}$$

where ω_c once again denotes a characteristic radius for the filter; its *order*, n, measures the abruptness of the cut-off. Radial sections of these filters are shown in Figure 7.16.

7.2.2 Digital simulation of analog filters

We next investigate the *realization* of filters, that is to say, methods for obtaining the filtering effect digitally. This problem is, of course, relevant for systems in general; accordingly, the term *system realization* or *system simulation* is used. An extensive body of theory is devoted to this topic; in what follows, we outline some very general principles and give a few illustrations.

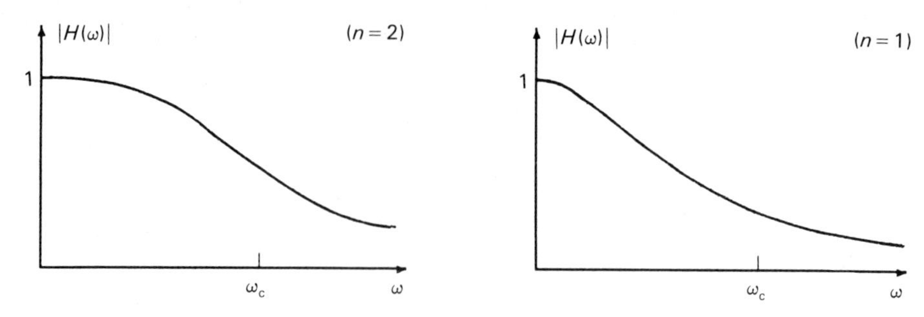

Figure 7.16 (a) Butterworth filter. (b) Exponential filter.

As we recall from Section 3.6, the system function $H(z)$ for a digital system is the z-transform of the impulse response h:

$$H(z) = \sum_{n=-\infty}^{\infty} h[n] z^{-n} \tag{7.16}$$

Again, we restrict ourselves to the complex unit circle; with a slight redefinition of H, we obtain

$$H(\omega) = \sum_{n=-\infty}^{\infty} h[n] \mathrm{e}^{-\mathrm{i}n\omega} \tag{7.17}$$

This specifies the filter function H as a Fourier series with coefficients $h[n]$ (or rather $h[-n]$, see equation (2.39)). According to equation (2.38), these coefficients are found from

$$h[n] = \frac{1}{2\pi} \int_{-\pi}^{\pi} H(\omega) \mathrm{e}^{\mathrm{i}n\omega} \, \mathrm{d}\omega \tag{7.18}$$

The digital filter to simulate H is now synthesized by choosing an appropriate N and calculating $h[-N], h[-N+1], \ldots, h[N]$. This implies that the sum $H_N(\omega) = \Sigma_{n=-N}^{N} h[n] \mathrm{e}^{-\mathrm{i}n\omega}$ will be a *least squares approximation* to $H(\omega)$, and the impulse response

$$h_N[n] = \begin{cases} h[n] & \text{for } n \in [-N, N] \\ 0 & \text{otherwise} \end{cases} \tag{7.19}$$

defines a system which, in the above sense, simulates (approximates) the analog system given by the filter function H.

The digital filter is thus described by

$$g[n] = \sum_{m} h_N[n-m] f[m] \tag{7.20}$$

where f is the filter input, and g its output.

EXAMPLE 7.5

The one-dimensional Hanning filter $H(\omega) = \frac{1}{2} + \frac{1}{2}\cos(\omega)$ corresponding to a cut-off frequency $\omega_c = \pi/2$ has Fourier series

$$H(\omega) = \tfrac{1}{2} + \tfrac{1}{4}e^{-i\omega} + \tfrac{1}{4}e^{i\omega}$$

with only three non-zero h-coefficients, $h[-1] = h[1] = \frac{1}{4}$ and $h[0] = \frac{1}{2}$. The filter is thus simulated digitally by

$$g[n] = \tfrac{1}{2}f[n] + \tfrac{1}{4}(f[n-1] + f[n+1])$$

and consists in the replacement of each signal value by $f[n]$, a weighted average in which the two neighbouring values are weighted by 50 per cent relative to the value itself.

Simulate a Hanning filter with five coefficients and the cut-off frequency $\omega_c = \pi/4$.

7.2.3 An example of a lowpass filter

Section 1.5 has already shown how image noise might be removed by binning pixels, that is, by forming averages over finite areas in the images. In the present context, this process entails the elimination of noise having spatial wavelengths down to about one pixel, that is, frequencies near the Nyquist frequency. This process must then somehow simulate a lowpass filter. This topic will now be touched upon quite briefly; once again, we simplify the notation by considering one-dimensional signals only.

Taking the average over a symmetrical area near each pixel corresponds, in the one-dimensional case, to

$$\bar{f}[n] = \frac{1}{2N+1}\sum_{m=-N}^{N} f[n-m]$$

that is, a convolution with an impulse response $h[n]$ which is a (shifted) rectangular signal:

$$h[n] = \frac{1}{2N+1} r_{2N+1}[n-N]$$

Thus, the filter function becomes

$$H(\omega) = \frac{1}{2N+1}\sum_{n=-N}^{N} e^{in\omega} = \frac{\sin(N+\frac{1}{2})\omega}{\sin(\frac{1}{2}\omega)} \tag{7.21}$$

as is easily seen by summing the quotient series, followed by multiplication by $e^{-i\omega/2}$ in both numerator and denominator (cf. Example 1.9).

EXAMPLE 7.6

Sketch this filter function and find an approximate value for its cut-off frequency. Discuss the significance of the periodicity of the filter function.

Next, let us compare this with the corresponding analog filter where the averaging about each point is described as a convolution with the rectangular signal

$$h(t) = \frac{1}{2T} r_{2T}(t - T) = \begin{cases} 1/2T & \text{for } |t| \leqslant T \\ 0 & \text{for } |t| > T \end{cases}$$

This filter has the system function (cf. Example 3.13 and Figure 3.8)

$$H_1(\omega) = \frac{\sin(T\omega)}{T\omega} \tag{7.22}$$

bearing a certain qualitative resemblance to the ideal lowpass filter in Figure 7.11a. Again, as an approximate measure of the filter bandwidth Ω, one may use the distance from $\omega = 0$ to the first zero, $\omega = \pi/T$, that is,

$$\Omega \approx \pi/T$$

expressing the relation between the size (T) of the averaging area if an approximate filter bandwidth Ω is to be achieved (cf. p. 119).

If the filter in equation (7.22) is squared, we obtain

$$H_2(\omega) = H_1(\omega)^2 = \frac{\sin^2(T\omega)}{(T\omega)^2} \tag{7.23}$$

providing an even more efficient rejection of the higher frequencies. This squaring in the frequency domain corresponds to a convolution in the time domain; the inverse Fourier transform of $H_2(\omega)$ is

$$h_2(t) = h_1 \star h_1(t) = \begin{cases} (T - |t|)/2T^2 & \text{for } |t| \leqslant T \\ 0 & \text{otherwise} \end{cases} \tag{7.24}$$

If, for instance, we wish to apply this filter to an image, its effect ought to be simulated by replacing each pixel value by a weighted average over the nearest neighbourhood, with weights decreasing linearly from the pixel in question.

The realization of this filter is achieved by the usual substitution of a sum for an integral, which after all is a least squares approximation (cf. Example 2.33).

For many simple filter simulations, the necessary calculations may be carried out directly, as the system function will contain only a few non-zero terms. For more complicated filters, a DFT may prove useful.

7.2.4 An example of a highpass filter

The *gradient* of an analog image $b(x, y)$ is the vector

$$\nabla b = \left(\frac{\partial f}{\partial x}, \frac{\partial f}{\partial y}\right) \tag{7.25}$$

which depends on position (x, y). The gradient is thus a vector field over the image – or, if one prefers, two separate images $\partial b/\partial x$ and $\partial b/\partial y$.

If the gradient vector is inserted at its position (x, y) in the image, it points in the direction where b changes most rapidly, and its length measures the change per unit length in that direction.

For instance, the gradient is large where an edge or a contour is present, and small where no significant intensity variations are encountered. With the input image b and the output $\partial b/\partial x$ (or $\partial b/\partial y$), a highpass filter is defined. We attempt to describe and simulate this filtering effect.

The system function H was introduced by means of its amplification of exponential (harmonic) signals (cf. Section 3.5.2). In the image case, we should consider $b_{uv}(x, y) = e^{iux}e^{ivy}$, with gradient

$$(\partial b_{uv}/\partial x, \partial b_{uv}/\partial y) = iub_{uv}, ivb_{uv}) \tag{7.26}$$

For the two gradient filters $\partial/\partial x$ and $\partial/\partial y$ taken separately, the amplification is thus iu and iv, respectively, that is,

$$H_x(u, v) = iu \qquad \text{and} \qquad H_y(u, v) = iv \tag{7.27}$$

Fourier analysis gives (cf. Example 2.39)

$$iu = \sum_{k \neq 0} \frac{(-1)^{k+1}}{k} e^{iku}$$

The coefficients of the two-dimensional Fourier series $H_x(u, v) = iu = \Sigma_{k,l}\, h_{kl}e^{iku}e^{ilv}$ thus become

$$h_{kl} = \begin{cases} 0 & \text{if } l \neq 0 \\ 0 & k = 0, l = 0 \\ (-1)^{k+1}/k & \text{otherwise} \end{cases} \tag{7.28}$$

The simulation of a gradient along the x-axis, using a digital 5×3 filter designed according to the guidelines described here, results in

0	0	0	0	0
$\frac{1}{2}$	-1	0	1	$-\frac{1}{2}$
0	0	0	0	0

Note the resemblance with the principles for recognition of simple objects (Section 5.3.1). The masks encountered there were, however, chosen according to more or less intuitive criteria, whereas those constructed now obey a certain optimality criterion – the resulting digital filter functions are least squares approximations to the ideal analog filters. A couple of examples follow.

EXAMPLE 7.7

Find a 3×3 filter simulating the effect of forming the gradient in the direction given by the vector (2, 1).

Hint: This gradient equals $2\partial b/\partial x + \partial b/\partial y$.

EXAMPLE 7.8

The Laplacian (operator) ∇^2 is defined by

$$\nabla^2 b(x, y) = \frac{\partial^2 b}{\partial x^2} + \frac{\partial^2 b}{\partial y^2}$$

Find the Laplacian system function $H(u, v)$ and show that the simulating filter is

0	2	0
2	$-2\pi^2/3$	2
0	2	0

In the following subsection, a new type of optimal filter is described.

7.2.5 Wiener filtering

One of the most frequently occurring filtering problems is the extraction of a signal from noise. Typically, a signal s is observed superimposed upon an (additive) noisy background b so that s is to be reconstructed from $s + b$ (cf. Figure 5.3). As mentioned, the noise is normally removed by taking simple or weighted averages; for images, this filtering might take place over square or circular image segments. The larger the segments, the more efficient the noise elimination – but at the same time, the signal itself is smoothed, possibly beyond recognition.

For given signal and noise properties, it is generally feasible to choose an optimal filter which removes as much as possible of the noise and retains as much as possible of the signal. The properties referred to are, understandably, associated with the *correlation* circumstances (Section 5.4) of the signal and the noise. A very popular filtering principle is due to the American mathematician Norbert Wiener.[1]

[1] Norbert Wiener, 1894–1964

Wiener filtering is based upon an a priori knowledge of signal and noise which is limited to the cross- and autocorrelation properties.

In the previous section, we simulated a filter optimally by approximating the filter function. The Wiener filter aims to approximate the signal itself. If the Wiener filter is presented with the sum $f = s + b$ of the signal and the noise as input, the output is

$$\bar{s} = h \star (s + b) = h \star f \tag{7.29}$$

where the optimality criterion for choosing h is a minimum signal distance $|s - \bar{s}|$.

To evaluate h we note that $\bar{s}$ appears as a linear combination of the time-translated versions of f; in digital notation, the signals are of the form $T_k f = f[n-k]$, where n denotes digital time, and k enumerates the translated signals. The way towards a solution is once again led by the *projection*, so as to make $s - \bar{s}$ orthogonal to all signals $T_k f$:

$$(s - \bar{s}) \cdot f[n-k] = 0 \qquad \text{for all } k \tag{7.30}$$

(cf. p. 58, especially equation (2.26)).

The reader is reminded that a scalar product of the form $f \cdot T_n g$ is called a correlation signal (denoted below by the letter c). Equation (7.30) now becomes

$$c_{fs}[n] = c_{f\bar{s}}[n]$$

expressing that the input f must correlate identically with the signal s and the filtered signal $\bar{s}$. From this and equation (7.29), we obtain

$$c_{fs} = h \star c_{ff}$$

from which h can easily be derived by Fourier transformation:

$$H(\omega) = \frac{C_{fs}}{C_{ff}} \tag{7.31}$$

Thus, in the frequency domain, the Wiener filter is given by the ratio between the Fourier transforms C_{fs} and C_{ff} of the respective correlation signals, where $C_{ff} = |F(\omega)|^2$ was earlier denoted Φ_f, the power spectrum of f.

Using the notation from pp. 152–6, then,

$$H(\omega) = \frac{|S(\omega)|^2}{|S(\omega)|^2 + |B(\omega)|^2} = \frac{\Phi_s^2}{\Phi_s^2 + \sigma^2} \tag{7.32}$$

if the noise is assumed to be white, to be of zero mean, and to have variance (mean power) σ^2.

EXAMPLE 7.9

A signal s has power spectrum $|S(\omega)|^2 = |\omega|^{-2}$ and is corrupted by additive white noise of mean power σ^2. Find the Wiener filter for this problem.

7.3 Image restoration

As mentioned earlier, the boundaries between the different image improvement schemes are somewhat diffuse. *Image restoration* is a special filtering technique which corrects for another, undesired, filtering mechanism having already been in play. This procedure is also known as *inverse filtering*, even if there is no difference of principle between direct and inverse filtering.

7.3.1 Inverse filtering

When restoring a given deteriorated image, we shall assume that the deteriorating mechanism has filter properties – it should be a *system* in the sense of Section 3.4: a (linear) transformation, the effect of which is independent (spatially invariant) of position across the image area. This assumption quite often holds in practice, although it is not hard to find several counter-examples. Many optical irregularities (coma, astigmatism, aberration, and so on) have varying properties across the image area in question; these variations, however, may often be considered constant, if sufficiently small areas are investigated.

If the above requirement is satisfied, the relation between input f and output g is

$$g = h \star f \tag{7.33}$$

where $h = PSF$ is the point spread function responsible for the deterioration. The given (blurred) image is g, and the task is to reconstruct the image f before the blurring.

We have already encountered several examples of image-blurring mechanisms with system properties (Figures 3.15 and 3.16, as well as Figures 7.3 and 7.4). We shall now develop various methods allowing at least a partial correction for the image deterioration just described. To simplify notation, only one-dimensional signals are considered.

After Fourier transformation, equation (7.33) is replaced by $G(\omega) = H(\omega)F(\omega)$, whence

$$F(\omega) = G(\omega)/H(\omega) \tag{7.34}$$

This relation is the formal solution of the inverse filtering problem. The reconstructed signal f is determined by its Fourier transform F, which in turn is obtained from the system function H and the Fourier transform G of the blurred signal. The system function of the inverse filter is, thus,

$$H_{\text{inv}} = 1/H(\omega) \tag{7.35}$$

Since the effect of the blur is described by a convolution, the inverse process – the reconstruction of the signal before blurring – is called a *deconvolution*.

Unfortunately, the above direct method turns out to be useless in practice. Not unexpectedly, the reason turns out to be noise and/or incomplete knowledge of the system function H.

If, for example, the signal is overlaid with noise b besides the degrading filter h, the relation between input f and output g becomes

$$g = h \star f + b \tag{7.36}$$

modifying equation (7.35) into

$$F(\omega) = \frac{G(\omega) - B(\omega)}{H(\omega)} \tag{7.37}$$

In general, the noise may very well possess large components at high frequencies ω, while g and h normally will be dominated by low-frequency components. For large ω-values, we have $G(\omega) \simeq 0$ and $H(\omega) \simeq 0$. The latter relation has, as evidenced by equation (7.37), devastating effect, since $F(\omega) \approx -B(\omega)/H(\omega)$, where both the numerator is large and the denominator small (in absolute terms). This qualitative argument shows that the noise will be amplified considerably in the inverse filtering, and the directly reconstructed signal will be worthless. The same conclusion is obtained by noting that, since (lowpass) filtering removes noise, inverse filtering must add noise. As is evident, an incompletely known function h may give rise to the same effect, since small, erroneous, values of $H(\omega)$ will imply large values of $F(\omega)$.

The image sequence in Figure 7.17 demonstrates the significance of the two sources of error: additive noise and an incorrect PSF. The original image (Figure 7.17a) has been blurred using a Gaussian PSF (cf. equation (7.39)), with Figure 7.17b as the result. The remaining images are deconvolutions of Figure 7.17b, where additive noise has been introduced before the deconvolution; moreover, a slightly inexact PSF has been employed. The first number is the signal/noise ratio for the noisy image before deconvolution, measured as the energy $|g|^2$ in the degraded image (Figure 7.17b) divided by the energy of the added noise image, $|b|^2$. The second number is the relative deviation between the PSF employed and the correct one (PSF_0), measured as $|PSF - PSF_0|^2/|PSF_0|^2$.

As will be evident, even relatively modest disturbances can play havoc with the direct deconvolution. We now attempt to find a way out of this predicament.

7.3.2 Determination of the point spread function

In order to carry out a meaningful image-restoration process, one must first of all know the point spread function $h(x, y)$ sufficiently well; alternatively, the system function $H(u, v)$, the two-dimensional Fourier transform of the PSF, might be specified. This condition is, however, rarely satisfied! If one is presented with a blurred image and nothing else, it will often be impossible to derive or read off the PSF directly.

(*a*) *Original image* (*b*) *Degraded image*

(*c*) *100:1, 0.01* (*d*) *10:1, 0.1*

(*e*) *100:, 0.2* (*f*) *10:1, 0.2*

Figure 7.17 Inverse filtering.

Even if a suitable point source (the image of which would be the PSF sought) is present in the scene being imaged, it will not exist in isolation, apart from the other objects, or even be located upon a constant background. The image of a possible point source will normally be superimposed upon undesirable contributions from other image areas – not to mention the image noise – and therefore unamenable to direct evaluation.

However, it is worth noting that several indirect methods exist for the evaluation of a PSF. For instance, an arbitrary object may serve our purpose, provided both the 'true' and the 'blurred' object are available (and at a satisfactory signal/noise ratio). If the true object is denoted b_0 and the blurred one b, we have $b = h \star b_0$, so that h may be determined from an inverse Fourier transformation of

$$H(\omega) = B(\omega)/B_0(\omega) \tag{7.38}$$

where B_0 and B are the Fourier transforms of the respective objects. This method has been utilized in Figure 7.18, in which Figure 7.18a mistakenly was photographed out of focus. But a renewed exposure with correct focusing (Figure 7.18b) allowed the PSF of the original image to be determined from the two versions of the object in the upper left-hand corner. Figure 7.18c shows the result of the inverse filtering of Figure 7.18a with the PSF obtained in this way.

Commonly employed reference objects in the determination of the PSF are straight lines, edges, corners, rectangles, and so on. In this context, terms such as *line spread function* apply. A substantial amount of work has been devoted to methods allowing a direct estimation of the PSF starting from these indirect characterizations of the blurring mechanism in an image.

The knowledge of the PSF of a given system is not necessarily derived from observations of point sources or other reference objects. Quite often, it stems from physical insight or previous experience concerning the workings of the imaging system; this knowledge will often be laid down in the form of an analytical expression, that is, a function $h = h(x, y)$.

A very popular PSF is the *Gaussian* (cf. equation (7.39) below), that is, a two-dimensional normal distribution. The Gaussian PSF is particularly justified where

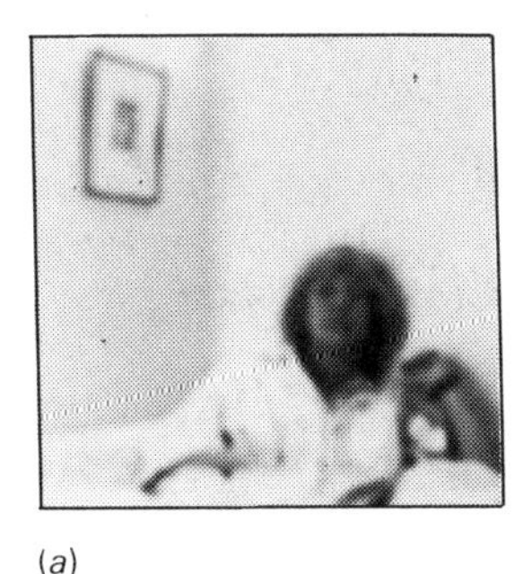

(a)

(b)

(c)

Figure 7.18 PSF determination by means of reference object.

the blurring mechanism may be viewed as a superposition of many independent small effects. This result derives from the *central limit theorem* (cf. Appendix A). Here, the PSF is considered as a two-dimensional probability distribution in the sense that $h[m\,n]$ is the probability that, due to the blurring, a photon is shifted m pixels horizontally and n pixels vertically away from the correct position in the image.

7.3.3 Interactive image restoration

Until this day, it has proved practically impossible to formulate objective criteria for *image quality*. For instance, computers remain unable to distinguish between 'sharp' and 'blurred' images. The removal of image blur, therefore, is often an *interactive* process in which the user constantly has to decide whether a result is acceptable. This decision must often be made as a comparison between several possible solutions to the same deconvolution problem, a subjective quality criterion being applied. Quite often, the solutions will be characterized by one or more *parameters*, the choice of which corresponds to a single and unique image reconstruction.

As a simple example of such an interactive image restoration process, see Figure 7.19. Figure 7.19a shows a blurred 550×400 digital image, but with an adequately large signal/noise ratio. We now assume the point spread function to be Gaussian with circular symmetry:

$$h(x, y) = \frac{1}{2\pi\sigma^2}\, e^{-(x^2+y^2)/2\sigma^2} \tag{7.39}$$

where the variance σ^2 is regarded as a free parameter. In other words, we apply the method of trial and error to various PSFs of different radii. Figures 7.19b–7.19f display the results; the current value for σ (in pixels) is given below each image. Most people will agree that Figure 7.19d is the most satisfactory reconstruction. If the blurring mechanism is *known* to be Gaussian, with PSF as in equation (7.39), it is probable that its radius is close to $\sigma = 3$.

The method is likely to yield acceptable results even if the blurring mechanism is unknown. As stated in the previous subsection, the Gaussian distribution is often an excellent approximation to the correct PSF, in particular where a physical justification exists.

We are now ready to become acquainted with a number of specific methods for image restoration. It will be assumed throughout that the PSF or system function is known and, not least, specified to a sufficient degree of accuracy.

7.3.4 Wiener deconvolution

Since the deconvolution using the inverse filter (equation (7.35)) introduces extra noise, one is tempted to consider a procedure where this noise is simply removed

(a) Original image
(b) $\sigma = 1$
(c) $\sigma = 2$
(d) $\sigma = 3$
(e) $\sigma = 4$
(f) $\sigma = 5$

Figure 7.19 Interactive determination of the PSF.

afterwards, for example, by means of a Wiener filter. The complete procedure is known as *Wiener deconvolution* and consists of a deconvolution followed by Wiener filtering. Finally, we investigate methods where the inevitable noise is introduced gradually, so to speak, and where the need for filtering will not become urgent.

Let us investigate the behaviour of the Fourier transforms of the signals. The given signal g has Fourier transform $G = HF + B$, with notation as previously. On passing the inverse filter $H_{\text{inv}} = 1/H$, the output transform becomes $G/H = F + B/H$, where B/H represents the inversely filtered (and thus increased) noise. This sum serves as input to the Wiener filter which, with the new noise contribution, becomes

$$H_{\text{W}}(\omega) = \frac{|F(\omega)|^2}{|F(\omega)|^2 + |B(\omega)|^2/|H(\omega)|^2}$$

When composing the two systems (cf. Section 3.1.1), the resulting deconvolution system emerges as follows:

$$H_{\text{W,inv}}(\omega) = H_{\text{W}}H_{\text{inv}} = \frac{1}{H(\omega)}\,\frac{|F(\omega)|^2}{|F(\omega)|^2 + |B(\omega)|^2/|H(\omega)|^2}$$

cf. equation (7.32). To summarize, the Wiener deconvolution (often misleadingly referred to as the 'Wiener filter', an ambiguous term), has the system function

$$H_{\text{W,inv}}(\omega) = \frac{H(\omega)^*}{|H(\omega)|^2 + \gamma|B(\omega)|^2/|F(\omega)|^2} \qquad (7.40)$$

where the factor γ introduced in the denominator should be unity. If, however, this factor is allowed to vary, we speak of a *parametric* Wiener filter for deconvolution. The value $\gamma = 0$ has the same effect as $B = 0$ and corresponds to the simple inverse filter. Other values yield non-optimal filters, which, however, may possess other convenient qualities. This method is suitable for interactive image restoration.

The result of a typical Wiener deconvolution is shown in Figure 7.20. Note that

Before

After

Figure 7.20 Wiener deconvolution.

the method assumes an a priori knowledge of the power spectrum $|F(\omega)|^2$. The reconstruction therefore concerns only the *phase spectrum* $F(\omega)/|F(\omega)|$ of f.

7.3.5 Constrained restoration

Even if the inverse filtering problem, in principle, is completely formulated in equation (7.36), this equation never provides a full specification in practice. Besides the causes cited for this (partly numerical difficulties, partly the enhanced noise), it often happens that $G(\omega)$ is known only through an insufficient number of samples $G(\omega_1), \ldots, G(\omega_N)$. This situation is well known in radio astronomy and in certain types of medical scanning where, in fact, the Fourier transform $G(u, v)$ is the directly observable quantity – not the $g(x, y)$ itself (cf. Section 3.7.4). In many cases, thus, the deconvolution problem is ill-defined and will possess many solutions consistent with the observations.

Accompanying the solution of a practical deconvolution problem, therefore, there should be some recipe facilitating the choice between the possible reconstructions. This recipe must somehow accommodate the notion of image quality, and the interactive image restoration plays a central role here. Due to the large degree of freedom as compared to the nominally very rigorous statement of the deconvolution problem, however, the task of inspecting all possible reconstructions of a given image becomes insurmountable. Instead, certain empirical criteria for image quality are applied.

The most frequently used methods today are based upon maximization of a certain *quality function* for the image. This means that one chooses a certain function which associates a number with each image; these values are taken to be measures of image quality. Among the various reconstructed images consistent with the given PSF and the available data, we single out the one which maximizes the quality function.

The maximization procedure should respect any other a priori information demanded by the reconstruction. This information is often expressed as *constraints*, that is, requirements that certain other functions of the images assume prescribed values. One typical constraint is the constancy of image energy and/or total intensity during each step of the reconstruction process.

A very common quality measure is the image *entropy*:

$$\mathcal{H} = -\sum_{m,n} b[m, n] \ln(b[m, n]) \tag{7.41}$$

In contrast to Chapter 4, the $\mathcal{H}$ defined here is not the entropy of the level distribution but that associated with the image itself, viewed as a two-dimensional probability distribution. Note that the maximization is indifferent to a possible normalization of b.

An intuitive rationale for employing $\mathcal{H}$ as a quality measure is the fact that $\mathcal{H}$ is known to quantify the concept of *smoothness*. As stated earlier, $\mathcal{H}$ is a maximum if the distribution in question is uniform (rectangular). In image restoration, the uniform distribution (that is, an image with constant intensity) is irrelevant as such. Nevertheless, the restored image may be qualitatively characterized as that exhibiting the largest degree of resemblance with the uniform distribution and which, at the same time, complies with the given data as well as the constraints.

The sequence in Figure 7.21 illustrates an application of the *maximum entropy* method. Under each image the computing time is given as a percentage of the total.

7.3.6 Removal of blur from uniform motion

To make the above principles a little more specific, we now construct a method allowing a solution to the problem indicated in Figure 7.3: removal of blur caused by motion of either the camera or the subject. We assume for simplicity that we are to reconstruct an entire image – in Figure 7.3 only a segment (the car) was blurred. Moreover, we assume the motion to be uniform during exposure. However, it does not constitute a further restriction to assume that the motion takes place exclusively in the x-direction of the image. If not, the image is rotated before restoration.

Our task is to improve the blurred digital image in Figure 7.22, consisting of 550×400 pixels. The blur is obviously only active in the horizontal direction. In other words, the PSF h is a function of the variable m only.

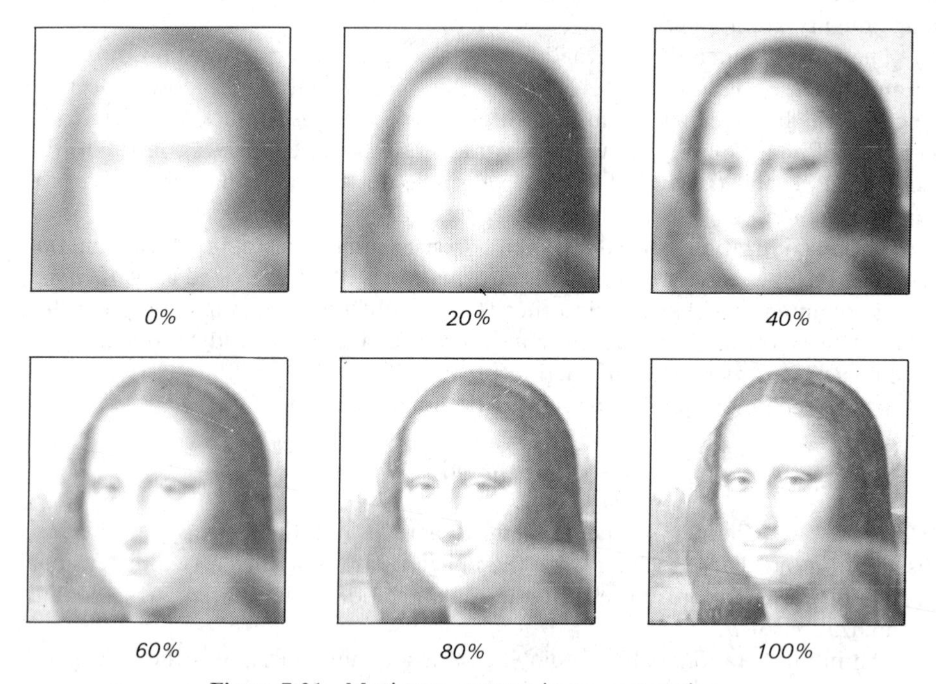

Figure 7.21 Maximum entropy image restoration.

Figure 7.22 Unidentified flying object.

$$h(=h[m,n]) = h[m]$$

and the point spread function must be rectangular:

$$h[m] = r_M[m] \tag{7.42}$$

a possible normalization factor of $1/M$ being omitted for convenience. The number M, measuring the extension of the blur in pixels, is found to be $M = 45$. (Try this for yourself.)

The deconvolution problem is now one-dimensional, and the image can be restored line by line. If a particular horizontal line (row) in the blurred image is denoted $g[m]$ and the corresponding deblurred row is $f[m]$, then

$$g[m] = \sum_k h[m-k]f[k] \tag{7.43}$$

This equation is easily solved by z-transformation. From Example 3.26 we first obtain

$$H(z) = \frac{1-z^{-M}}{1-z^{-1}}$$

so that the inverse filter is described by the system function

$$H_{\text{inv}}(z) = \frac{1}{H(z)} = \frac{1-z^{-1}}{1-z^{-M}} \tag{7.44}$$

Recognizing $1/(1-z^{-M})$ as the sum of an infinite quotient series,

$$\sum_{k=0}^{\infty} (z^{-M})^k = \frac{1}{1-z^{-M}}$$

the system function, written as a power series, becomes

$$H_{\text{inv}}(z) = \sum_{k=0}^{\infty} (z^{-kM} - z^{-kM-1}) \tag{7.45}$$

An inverse z-transformation next provides the impulse response for the deconvolution filter:

$$h_{\text{inv}}[m] = \begin{cases} 1 & m = kM \\ -1 & m = kM + 1 \\ 0 & \text{otherwise} \end{cases} \qquad (k \in [0, \infty]) \tag{7.46}$$

that is, the coefficients in the power series in equation (7.45). This filter is shown graphically in Figure 7.23.

EXAMPLE 7.10

Show directly that $h_{\text{inv}}[m] \star h[m] = \delta[m]$. Comment on the significance of the periodicities and the possible edge effects.

Figure 7.24 illustrates the effect of the deconvolution filter (equation (7.46)) applied to the image in Figure 7.22, taken line by line.

EXAMPLE 7.11

The method might be of interest for amateur astronomers photographing the sky with a CCD camera. If the telescope is fixed – or otherwise unable to follow the

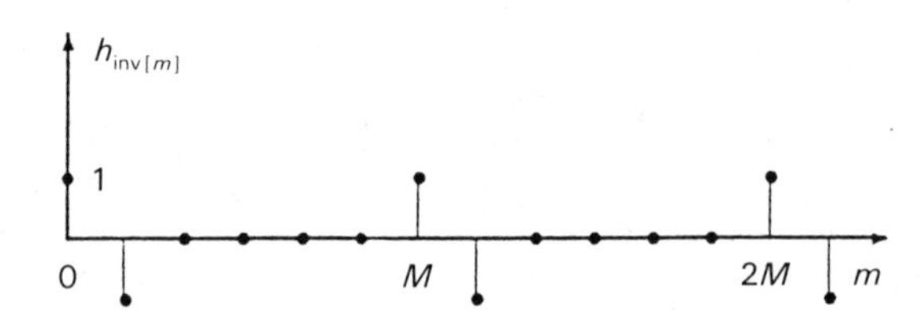

Figure 7.23 Filter for deconvolution of motion blur.

Figure 7.24 Correction for motion blur.

celestial sphere in its diurnal motion – all celestial objects will be smeared into small line segments. As the above considerations show, however, this problem is easily eliminated.

7.3.7 Deconvolution and moments

In this and the following subsection, we examine two deconvolution methods, each of which is of very broad design; in practice, however, they each apply to rather specific types of problem.

The first method involves a determination of the *moments* of the restored images, starting from the moments of the blurred image and those of the PSF. As mentioned in Appendix A, a probability distribution – one of the interpretations of a digital image – is completely characterized in terms of its moments, and an evaluation of the complete set of moments for the deconvolved image implies, in principle, its full characterization. In practice, things are not that easy, as the image noise limits the efficiency of the method.

Nevertheless, the method is excellent for recovering images of certain simple objects, the structure of which is given by a few moments (such as ovals, pear or heart shapes). In particular, the method is extremely useful in the context of *recognition* (cf. Section 5.6), notably where this task is performed by computer. Here, the correction for smearing may be essential, as illustrated in Figure 7.25.

Figure 7.25a shows a blurred image of a fruit, known, however, to be an avocado, a lemon, or a papaya (Figure 7.25b). We now develop a simple method, based upon the moments of these objects, allowing an identification of the object in question.

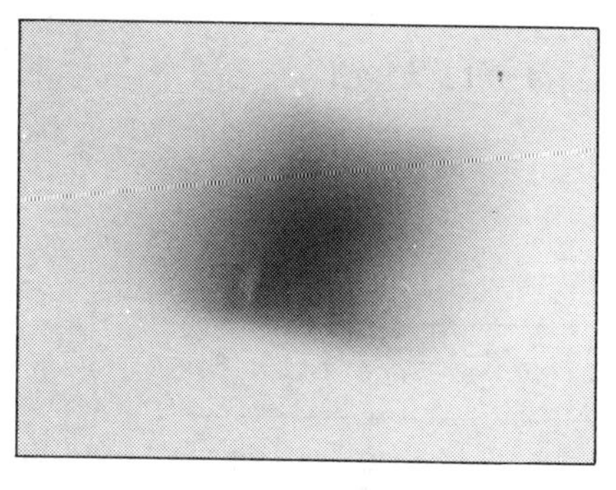

(*a*)

(*b*)

Figure 7.25 Blurred piece of fruit with three possible identifications.

The reader is reminded that the moment b_{kl} of the object $b = b[m, n]$ is defined by

$$b_{kl} = \sum_m \sum_n m^k n^l b[m, n] \tag{7.47}$$

As was shown in Section 5.6, a proper choice of coordinate system for each object ensures $b_{10} = b_{01} = b_{11} = 0$. We may also assume that b is normalized ($b_{00} = 1$). The latter assumption implies that b defines a two-dimensional probability distribution.

With these assumptions, the second-order moments for the three objects are as listed in Table 7.1. The moments of the PSF were determined from a small point, blurred together with the unidentified object.

Now for the identification. If the sharp object, the PSF, and the blurred object are called b, h, and c, respectively, we have the relation

$$c[m, n] = \sum_{m_1} \sum_{n_1} h[m - m_1, n - n_1] b[m_1, n_1] \tag{7.48}$$

which, as regards the moments, implies

$$\begin{aligned} c_{kl} &= \sum_m \sum_n m^k n^l c[m, n] \\ &= \sum_m \sum_n m^k n^l \sum_{m_1} \sum_{n_1} h[m - m_1, n - n_1] b[m_1, n_1] \\ &= \sum_{m_1} \sum_{n_1} \sum_m \sum_n m^k n^l h[m - m_1, n - n_1] b[m_1, n_1] \end{aligned}$$

In order to simplify the notation, we briefly omit one dimension, so that

$$c_k = \sum_{m_1} \sum_m m^k h[m - m_1] b[m_1] = \sum_{m_1} b[m_1] \sum_{m_2} (m_2 + m_1)^k h[m_2]$$

where the summation variable m has been replaced by $m_2 = m - m_1$. Next, $(m_2 + m_1)^k$ is expressed by means of the binomial theorem:

$$c_k = \sum_{m_1} b[m_1] \sum_{m_2} \sum_i \binom{k}{i} m_2^i m_1^{k-i} h[m_2]$$

Table 7.1 Moments for three fruits.

Object	*PSF*	*Avocado*	*Lemon*	*Papaya*	*?*
b_{20}	6	65	65	70	70
b_{11}	− 12	0	0	0	− 10
b_{02}	28	20	30	20	50

Interchanging again the order of summation leads to

$$c_k = \sum_i \binom{k}{i} \sum_{m_1} m_1^{k-i} b[m_1] \sum_{m_2} m_2^i h[m_2]$$

where the last two sums are recognized as the moments b_{k-i} and h_i associated with b and h, respectively. Restoring the two-dimensional notation, the total result thus becomes

$$c_{kl} = (h \star b)_{kl} = \sum_{i=0}^{k} \sum_{j=0}^{l} \binom{k}{i}\binom{l}{j} h_{ij} b_{k-i,l-j} \tag{7.49}$$

This fundamental relation mediates an evaluation of the moments for the corrected object from the moments of the given (blurred) object, if the PSF moments are known as well. The sequence of explicit formulae starts as follows:

$$\begin{aligned} c_{00} &= h_{00}b_{00} \\ c_{10} &= h_{00}b_{10} + h_{10}b_{00} \\ c_{01} &= h_{00}b_{01} + h_{01}b_{00} \\ c_{20} &= h_{00}b_{20} + 2h_{10}b_{10} + h_{20}b_{00} \\ c_{11} &= h_{00}b_{11} + h_{01}b_{10} + h_{10}b_{01} + h_{11}b_{00} \\ c_{02} &= h_{00}b_{02} + 2h_{01}b_{01} + h_{02}b_{00} \end{aligned} \tag{7.50}$$

and, as will be seen, the quantities b_{kl} may be found successively.

EXAMPLE 7.12

Through the action of a point spread function h, an image b has been blurred, resulting in the image c. The moments up to second order for the images c and h are measured as listed in the table below:

Moment no.	00	10	01	20	11	02
c	8	2	4	32	1	32
h	2	1	1	2	−4	0.5
b						

Fill in the table with the moments for the deconvolved image b.

EXAMPLE 7.13

Identify the unknown object in Figure 7.25a, given the information in Table 7.1.

Normally, the coordinate system is placed in the 'standard' position for c, so that $c_{01} = c_{10} = c_{11} = 0$; likewise, h is assumed to obey $h_{01} = h_{10} = h_{11} = 0$. This will be the case if h possesses circular symmetry (or is oriented as c). We assume, moreover, that $c_{00} = h_{00} = 1$ (normalization of image and PSF). If so,

$$b_{00} = 1 \qquad \text{and} \qquad b_{01} = b_{10} = 0$$
$$b_{20} = c_{20} - h_{20}, \qquad b_{11} = 0, \qquad b_{02} = c_{02} - h_{02} \tag{7.51}$$

showing that b is oriented as c. Equations (7.51) then give the relationship between the *invariant* moments of b and c.

EXAMPLE 7.14
Find the four deconvolved moments of third order (with the same assumptions as above).

EXAMPLE 7.15
An elliptical object (see, for example, Figure 3.14) has been blurred with a circular symmetric point spread function. The PSF is unknown, but a circular object b, known to be degraded by the same mechanism as the ellipse, has $b_{20} = b_{02} = 2$, and, for the blurred circle, the measurements $c_{20} = c_{02} = 5$ are obtained. For the blurred ellipse, the two invariant moments of second order are measured to be 5 and 4 (that is, the axis ratio is $\sqrt{5/4}$). Find the correct axis ratio for the elliptical object.

In principle, the method of moments for inverse filtering is completely general and allows a deconvolution of any blurred object. However, as equation (7.47) indicates, image errors or noise will contribute heavily to c_{kl}, especially if occurring in pixels far from $[0, 0]$, and if the order of the moment, $k + l$, is high. The procedure is therefore particularly suited to reconstruction of images (objects) known to possess a high degree of symmetry, so that it is well described in terms of a few low-order moments.

7.3.8 The CLEAN algorithm

The CLEAN algorithm was originally designed for the processing of radio-astronomical image data. The images in Figure 3.16 have been processed in this way.

As was the case for the deconvolution method of the previous subsection, CLEAN works best on certain objects. These are images consisting of a small number of blurred point sources (of different strengths) observed against a slowly varying background. Since *every* digital image is a sum of point sources of different strengths (cf. equation (3.8)) it is implicitly understood that the sources in question are sparse and reasonably well separated.

The procedure starts by identifying the pixel $[m_1, n_1]$ in a given blurred image $c[m, n]$, where c is a maximum:

$$c[m, n] \leqslant c[m_1, n_1] \qquad \text{for all } m, n$$

The working hypothesis of the method is as follows. In pixel $[m_1, n_1]$ in the rectified image b, there is a point source of strength b_1 contributing the image component $b_1\delta[m - m_1, n - n_1]$ to b and, hence, the component $b_1h_1[m, n] = b_1h[m - m_1, n - n_1]$ to c. On subtracting b_1h_1 from c, we obtain a new image c_1:

$$c_1[m, n] = c[m, n] - b_1h[m - m_1, n - n_1]$$

which has been 'cleaned' by the point source in question.

The problem is the determination of the strength b_1. Various criteria exist (see below); an appropriate strategy consists in the minimalization of the image distance $|c - b_1h_1|$ (cf. p. 161).

Next, the procedure is repeated for the image c_1: the maximum pixel $[m_2, n_2]$ is located, and a point source of strength b_2 is placed here. After blurring and subtraction, we are left with

$$\begin{aligned} c_2[m, n] &= c_1[m, n] - b_2h[m - m_2, n - n_2] \\ &= c[m, n] - (b_1h[m - m_1, n - n_2] + b_2h[m - m_2, n - n_2]) \end{aligned}$$

The procedure is repeated until the only remaining image values all fall below a certain prescribed limit, fixed for example from consideration of the image noise or from the strengths of the point sources identified so far. The reconstructed image is then

$$b[m, n] = b_1\delta[m - m_1, n - n_1] + \cdots + b_N\delta[m - m_N, n - n_N] \qquad (7.52)$$

plus a residual image, $c_{N+1}[m, n]$.

Obviously, the method can lead to peculiarities such as negative intensities or artificial point sources. These problems can to some extent be circumvented. A commonly adopted provision consists in a more 'gentle' point source subtraction, for example by removing only 50 per cent of the derived contribution in each step.

CLEAN has a feature in common with several other image-restoration algorithms. During its development, it has been endowed with a number of extra 'magic' principles, the introduction of which is almost impossible to justify in a rigorous manner; nevertheless, they are indispensable ingredients in the solution of many problems. Such principles add to the elevation of image processing from workmanship to art....

Figure 7.26 shows the CLEAN algorithm in operation; the number of iterations is indicated under each resulting image. For the sake of illustration, the extracted point sources have been added successively to the CLEANed image.

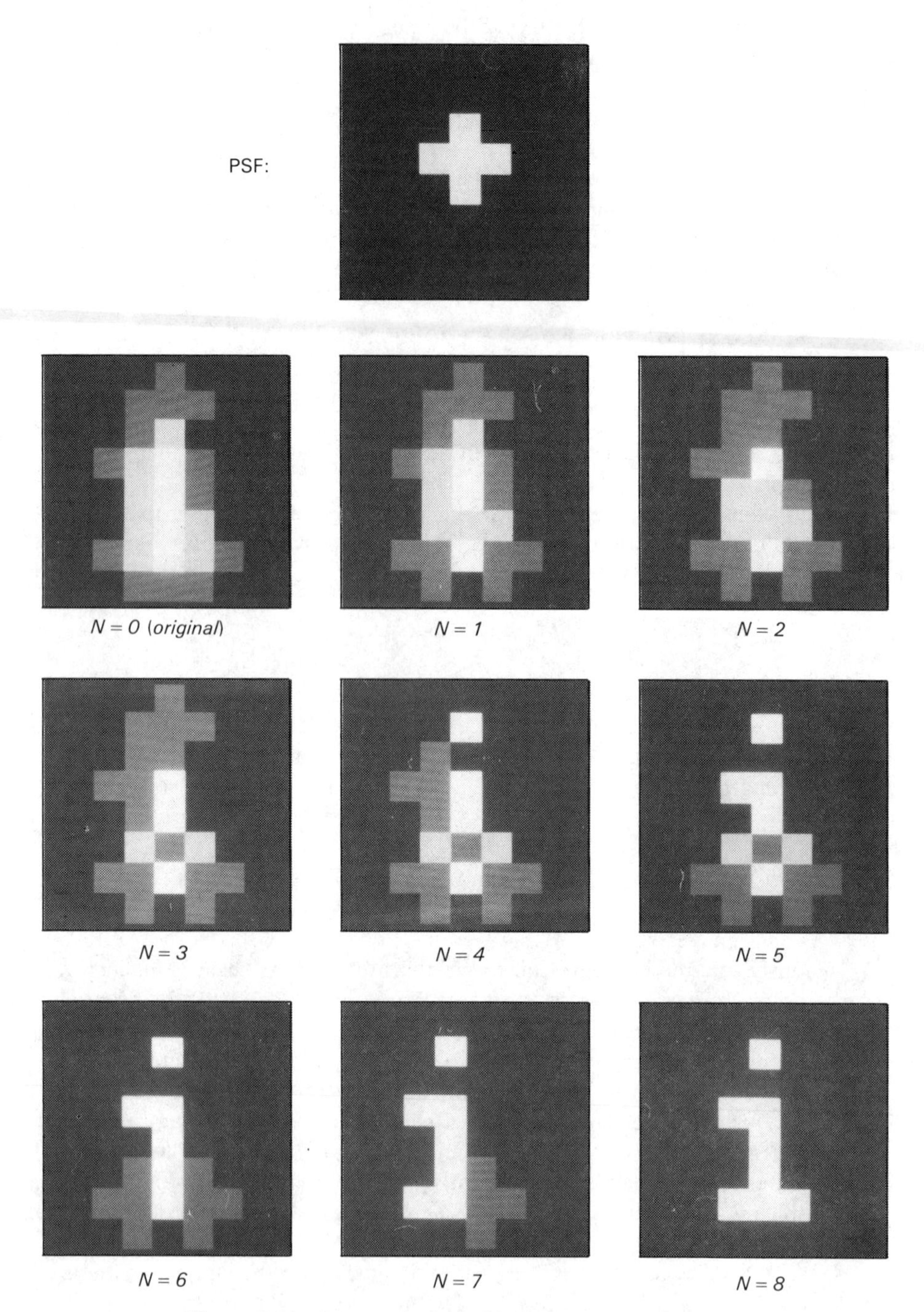

Figure 7.26 Deconvolution with CLEAN (schematic).

7.3.9 The Richardson–Lucy algorithm

As mentioned, the problem with simple inverse filtering was the possible amplification of the image noise to an unacceptable level. Consequently, this noise had to be removed, for instance by means of a Wiener filter. For the sake of clarity, this principle is illustrated once again in Figure 7.27.

The original, blurred, image is Figure 7.17b, improved by means of inverse filtering; the result is shown in Figure 7.27a, which is identical to Figure 7.17d. Even if this image represents the mathematically correct deconvolution of the original image (including its noise), the image quality is lost: the filtering from Figure 7.27a to Figure 7.27b is only partially successful in repairing the damage. One might imagine a less abrupt process by performing the restoration in a number of steps, each of which introduces only a small amount of noise, and where the original image quality (its smoothness) is preserved.

This is precisely the philosophy behind the maximum entropy method (p. 209). Here, the blurred image is altered gradually while the image entropy $\mathcal{H}$ (equation (7.41)) controls the smoothness. The resulting image thus represents a satisfactory compromise between the smoothness in the original image and the mathematics of the inversely filtered image.

Another iterative method which represents a compromise between 'quality' and 'accuracy' was introduced in the image processing literature by W. H. Richardson (1972) and L. B. Lucy (1974) – see Appendix D. The method is based on the Bayesian probability principles mentioned in Appendix A.

In this scheme, the addition of noise to the blurred image is considered part of the degradation. Since the blurring may be regarded as a random shift of the individual photons, hence a statistical process, this concept is by no means unnatural.

Thus, given a sharp and noiseless image b (the latter property may be expressed in the pixel values being *numbers*, not stochastic variables), the blurring might, in

(a) Inverse-filtered image

(b) Inverse and Wiener filtered

Figure 7.27 Deconvolution with noise filtering.

principle, produce any other image c. These results are, however, far from being equally likely.

The various resulting images c' may thus be assigned a (conditional) probability $P(c' \mid b)$, and the 'correct' result c of the blurring maximizes this probability, that is, $P(c' \mid b) \leqslant P(c \mid b)$. It seems reasonable to express the image-restoration principle along the same lines: Among the possible reconstructions b' of a given image c, we pick the one maximizing the probability $P(b' \mid c)$, that is,

$$P(b' \mid c) \leqslant P(b \mid c)$$

We now make these considerations slightly more specific. Once again, the notation is simplified to one-dimensional 'images'. The point spread function $h[n]$ is viewed as the probability of finding a photon in pixel n, given the correct position $n = 0$. The invariance (the system property) implies that the probability of finding a photon in pixel n, given the correct position k, is

$$P(n \mid k) = h[n-k]$$

With the original image $b[n]$ (which we assume to be statistically normalized, that is, $\Sigma b[n] = 1$), the total probability distribution for the 'blurred photons' becomes $\Sigma_k P(n \mid k) b[k] = \Sigma_k h[n-k] b[k]$: the one defined by the blurred image.

If the 'inverse', conditional, probability is denoted $Q(k \mid n)$, which is the probability that the photon in pixel k originated in the (correct) position n, we have, according to Bayes's rule (equation (A.7) in Appendix A)

$$Q(k \mid n) = \frac{h[n-k]\, b[k]}{c[n]} \tag{7.53}$$

Summation over n yields

$$\sum_n Q(k \mid n) c[n] = \sum_n h[n-k]\, b[k] = b[k] \sum_n h[n-k]$$

Since h is normalized, b is determined by

$$b[k] = \sum_n Q(k \mid n) c[n] \tag{7.54}$$

$Q(k \mid n)$ being unknown, the solution is not quite to hand yet. However, equations (7.53) and (7.54) together suggest an iterative method for the evaluation of b.

Start with a proposed solution b_0 and construct the corresponding $c_0 = h \star b_0$. From equation (7.53), an approximate determination of Q results, namely, $Q_0(k \mid n) = h[n-k]\, b_0[k] / c_0[n]$, which is substituted into equation (7.54), yielding $b_1[k] = \Sigma_n Q_0(k \mid n) c[n]$ (here, the given image c is used).

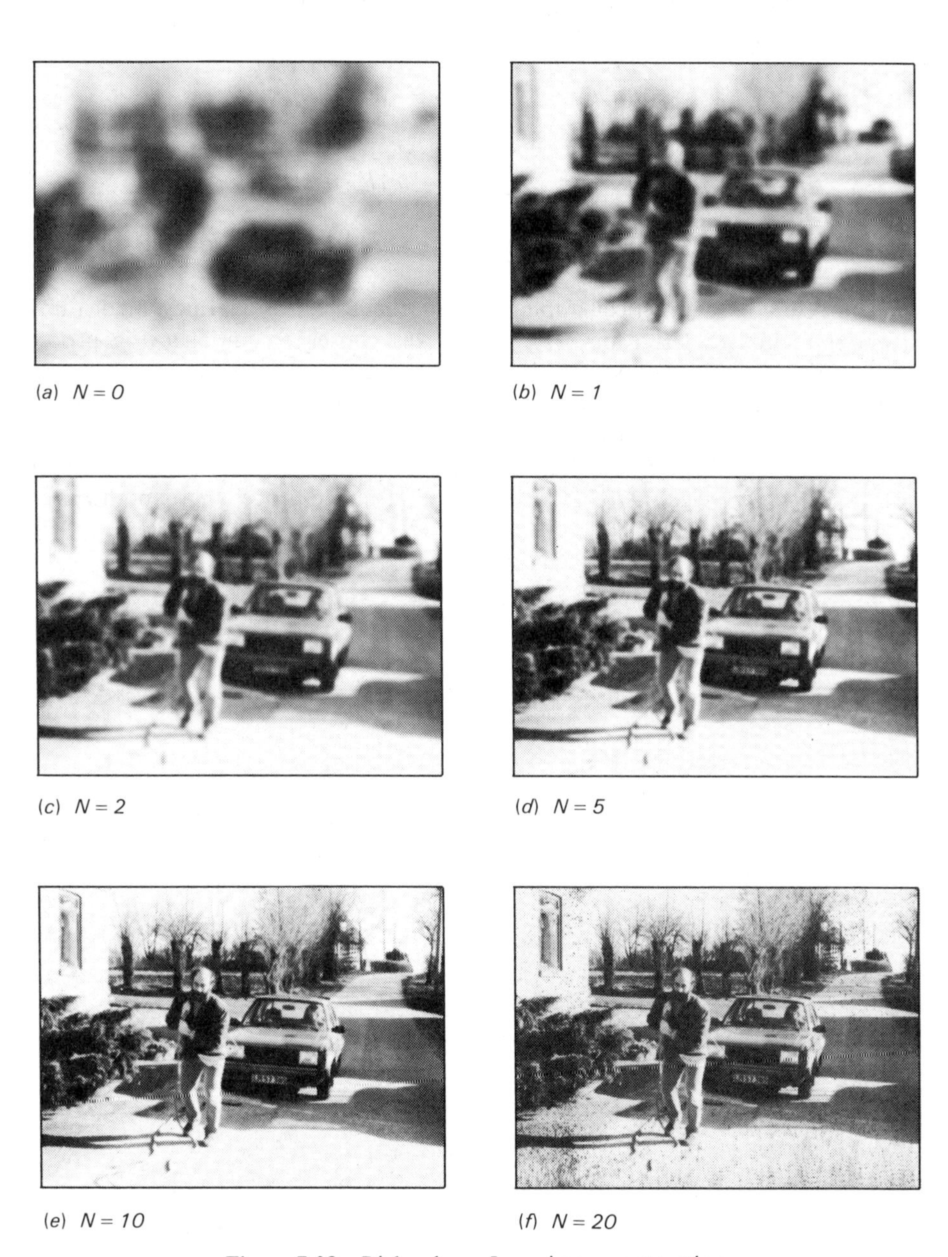

Figure 7.28 Richardson–Lucy image restoration.

The successive determinations of $b_1, b_2, \ldots$ lead to the following iterative scheme, where the two-dimensional notation is reinstated for good measure:

$$c_j[m,n] = \sum_k \sum_l h[m-k, n-l]\, b_j[k,l]$$

$$b_{j+1}[m,n]/b_j[m,n] = \sum_k \sum_l h[k-m, l-n]\, c[k,l]/c_j[k,l] \qquad (7.55)$$

It can be shown that the images b_j converge towards the most probable solution of the deconvolution problem.

The result of a practical application of the Richardson–Lucy algorithm is shown in Figure 7.28 (see p. 221). $b_0 = c$ was chosen to initialize the process. Characteristically, only a few iterations are necessary before the resulting image is satisfactory. Further iterations tend to lower the image quality, although the image converges towards the mathematically correct solution.

Finally, it should be noted that the efficiency of the method is greatly improved by an intelligent first guess b_0. Here, too, digital image processing is both art and science!

APPENDIX A

Basic Statistical Concepts

A.1 Experiment and outcome

The human description and understanding of the surrounding world is intimately linked with our ability to *predict* chains of events in given situations, that is, the ability to combine cause and effect. This is exploited in practice when certain circumstances are found again at a new time and/or a new place.

We are very often unable to predict imminent events, as the circumstances are too complicated or inadequately known. But very many situations exist where another kind of description applies – statistics. This description is based upon a *large* number of repetitions of the same circumstances.

EXAMPLE A.1
Before a die is cast, one is unable to predict the outcome. However, when a large number of these 'experiments' are performed, the number of 1s will approximately equal the number of 2s, 3s, and so on. The number of, say, 5s will be approximately one-sixth of the total number of trials, and the natural generalization is the statement that the ratio is *exactly* one-sixth in an infinite number of trials.

The concepts of *cause* and *effect* are, in the statistical description, replaced by *experiment* and *outcome*. The outcome of an experiment is unpredictable; its *frequency* in a large number of repetitions is, however, predictable.

We thus assume that the experiment has N different equiprobable outcomes (in the throw of a die, $N = 6$) which are equivalent with respect to the given circumstances. Each of the outcomes $u_1, \ldots, u_N$ is therefore assigned the probability

$$p_n = p(u_n) = 1/N \qquad \text{(A.1)}$$

On enumerating the outcomes, they may be replaced by the integers $1, 2, \ldots, N$. Each number $n \in [1, N]$ is thus assigned the probability $p_n = 1/N$, and the function p_n, $n \in [1, N]$, is called the *probability distribution* of the experiment. Since all the probabilities are equal, we speak of a *uniform* or *rectangular distribution* on the set of all possible outcomes, the *sample space* $[1, N]$. If we next consider a group consisting of k outcomes, we naturally assign the probability $p = k/N$ to this group. Such a 'composite' outcome is called an *event*. If the sample space is divided into M mutually exclusive events having probabilities p_m, we still have

$$\sum_m p_m = 1, \quad (p_m \geqslant 0) \tag{A.2}$$

EXAMPLE A.2

There is a probability of 1/52 that the uppermost card in a deck that has been shuffled is the 5 of spades. There is a probability of 13/52 = 1/4 that the uppermost card is a spade: The event 'spades' consists of 13 outcomes, each having a probability of 1/52.

An event is thus simply a subset H of the sample space $[1, N]$. If H consists of the outcomes $u_1, \ldots, u_K$, its probability becomes

$$p(H) = \sum_{n=1}^{K} p_n = \frac{K}{N}$$

or, in an obvious notation,

$$p(H) = \sum_{n \in H} p_n = \frac{\#(H)}{N} \tag{A.3}$$

where $\#(H)$ denotes the number of outcomes in H.

EXAMPLE A.3

For the 52 cards in a deck, the ace of spades may be assigned the number 1, the 2 of spades, 2, and so on; next, hearts, diamonds, and clubs are assigned numbers. The outcome '7 of diamonds' then becomes 33, and 'king of clubs' becomes 52. The event 'red card' equals $R = [14, 39]$, 'black card' equals the complementary set $S = [1, 13] \cup [40, 52]$. We have

$$p(R) = p(S) = 26/52 = 1/2$$

What is the probability of a randomly chosen card being

(a) a spade, (b) an ace and/or a spade, (c) *either* an ace *or* a spade?

A.2 Stochastic variables and probability distributions

In order to be able to enumerate different events or, at least, to associate them with numbers, we use functions (digital signals, if you like!) in $[1, N]$. Inasmuch as this interval is considered as consisting of outcomes, such a function is termed a *stochastic variable*. If X is a stochastic variable and $x_1, \ldots, x_M$ its range of values, we may determine the probability of X assuming the value x_m. The outcomes $n \in [1, N]$ for which $\mathsf{X}[n] = x_m$ form an event H_m, denoted $(\mathsf{X} = x_m)$, for which

$$q_m = p(H_m) = p(\mathsf{X} = x_m) \tag{A.4}$$

By including all the values x_m assumed by X, we obtain a set of events $H_1, \ldots, H_M$ which make up a *partition* of $[1, N]$ (each number $n \in [1, N]$ belongs to one, and only one, of the events H_m). This means that the qs form a probability distribution:

$$\sum_{m=1}^{M} q_m = 1$$

called the *distribution of the stochastic variable* X. The original distribution may therefore be considered as the distribution of the variable n given by $\mathsf{n}[m] = m$. Here, each outcome H_m consists of the outcome m only.

If we proceed further and consider $[1, M]$ instead of $x_1, \ldots, x_M$ as the set of outcomes for X, where necessarily $M \leqslant N$, we have constructed a probability distribution q_m on this interval.

EXAMPLE A.4

The stochastic variable X on the interval $[1, 5]$ given by

$$\mathsf{X}[n] = \begin{cases} 1 & n \in [1, 2] \\ 2 & n \in [3, 5] \end{cases}$$

has the distribution $q_1 = 2/5$, $q_2 = 3/5$.

EXAMPLE A.5

A stochastic variable X is defined on the interval $[1, 10]$ as follows:

$$\mathsf{X}[n] = \begin{cases} 0 & n \in [1, 3] \\ n - 3 & n \in [4, 6] \\ n - 7 & n \in [7, 10] \end{cases}$$

For this stochastic variable, find

(a) its induced partition of $[1, 10]$ and
(b) its distribution.

If we wish to discuss stochastic variables and partitions of the new sample space $[1, M]$, no new concepts need be defined. A stochastic variable Y is still a function on $[1, M]$; let us assume its sample space to be $y_1, \ldots, y_L$ or simply $[1, L]$ $(L \leqslant M)$. Its *distribution* r_l, $l \in [1, L]$, is also in this case defined by the probabilities

$$r_l = q(\mathsf{Y} = l) \tag{A.5}$$

of Y assuming the value l, that is, of the events $(\mathsf{Y} = l)$, $l \in [1, L]$.

It is, however, easy to see that r_l is precisely the probability of $\mathsf{Y} \circ \mathsf{X}[n] = l$. Here, $\mathsf{Y} \circ \mathsf{X}$ denotes the *composite* stochastic variable on $[1, N]$ given by $\mathsf{Y} \circ \mathsf{X}[n] = \mathsf{Y}[\mathsf{X}[n]]$. *The distribution of* Y (defined for the probability distribution q_m) *is therefore obtained as the distribution of* $\mathsf{Y} \circ \mathsf{X}$ (on the rectangular distribution $p_n = 1/N$).

In general, each stochastic variable is associated with two distributions: an 'original' and an *induced* distribution.

EXAMPLE A.6
Here, the probability distribution

$$q_n = n/10, \qquad n \in [1, 4]$$

is given. The stochastic variable

$$\mathsf{Y}[n] = \begin{cases} 1 & n \in [1, 2] \\ 2 & n \in [3, 4] \end{cases}$$

has the distribution $r_1 = 1/10 + 2/10 = 3/10$, $r_2 = 3/10 + 4/10 = 7/10$.

Find the distribution of the stochastic variables

(a) $\mathsf{Y}[n] = n^2$ (b) $\mathsf{Y}[n] = (n-3)^2$ (c) $\mathsf{Y}[n] = 2n^3 - 15n^2 + 37n$.

A.3 Conditional probabilities and Bayes's rule

An important concept appearing in conjunction with partitions of the sample space is the so-called *conditional probability distribution*. If the M events $H_1, \ldots, H_M$ constitute a partition of the sample space $[1, N]$, and H is an arbitrary event, then there will be a definite probability of both H and H_m occurring; this event is denoted $H \cap H_m$. We clearly have

$$\sum_{m=1}^{M} p(H \cap H_m) = p(H)$$

If the event H is known to occur with probability greater than 0, we obtain a new distribution of probabilities for H_m occurring as well:

$$p(H_m \mid H) = p(H_m \cap H)/p(H) \tag{A.6}$$

($p(H_m \mid H)$ is read as 'the probability of H_m *given* H').

Since $p(H_m \cap H) = p(H \mid H_m)p(H_m)$, we obtain

$$p(H_m \mid H) = \frac{p(H \mid H_m)p(H_m)}{p(H)} \tag{A.7}$$

This important result is called Bayes's rule.

EXAMPLE A.7

On a rack K_1 there hang 4 green and 2 red hats; on K_2, 3 green hats and 5 red ones. I choose a rack and one of its hats at random. The probabilities are, with an obvious notation,

$$p(K_1) = p(K_2) = 1/2$$
$$p(G \mid K_1) = 2/3, \qquad p(R \mid K_1) = 1/3$$
$$p(G \mid K_2) = 3/8, \qquad p(R \mid K_2) = 5/8$$

$$\begin{aligned} p(G) &= p(G \mid K_1)p(K_1) + p(G \mid K_2)p(K_2) \\ &= (2/3)\cdot(1/2) + (3/8)\cdot(1/2) = 25/48 \\ p(R) &= p(R \mid K_1)p(K_1) + p(R \mid K_2)p(K_2) \\ &= (1/3)\cdot(1/2) + (5/8)\cdot(1/2) = 23/48 \end{aligned}$$

I find myself wearing a green hat. What is the probability of my having chosen rack 1?

Solution:

$$p(K_1 \mid G) = \frac{p(G \mid K_1)p(K_1)}{p(G)} = \frac{\frac{2}{3}\cdot\frac{1}{2}}{\frac{25}{48}} = \frac{16}{25}$$

If I am wearing a red hat, what is the probability that I chose rack 1?

A.4 Two-dimensional distributions

In image processing, a naturally occurring theme is two-dimensional probability distributions, that is, quantities p_{mn}, $m \in [1, M]$, $n \in [1, N]$, where the p_{mn}s are positive numbers obeying

$$\sum_{m=1}^{M} \sum_{n=1}^{N} p_{mn} = 1$$

If a digital image $b[m,n]$ is normalized in the statistical sense, that is, divided by the total number of counts $b = \Sigma_{m,n}\, b[m,n]$, a probability distribution results. This distribution is an approximation to the ideal image distribution occurring in the case of an infinite number of counts – as in the uniform distribution arising on the interval $[1,6]$ for infinitely many throws of an unbiased die. Thus, $p_{mn} = b[m,n]/b$ may be regarded as the probability that an incoming photon is recorded at pixel $[m,n]$. Or one may see fit to consider the photon position as a two-dimensional stochastic variable $[\mathsf{m},\mathsf{n}]$ which, with probability p_{mn}, assumes the value $[m,n]$.

The *columns* $S_1, \ldots, S_M$ of a digital image constitute a partition of it. If we also consider the nth *row* R_n, we find

$$p(R_n) = \sum_{m=1}^{M} p_{mn}$$

and the conditional distribution

$$p(S_m \mid R_n) = p_{mn}/p(R_n), \qquad m \in [1, M]$$

If the various distributions resulting from different choices of $n \in [1, N]$ are in fact identical, that is, independent of n, then we may write $p_m = p_{mn}/p(R_n)$ instead, where p_m is the common probability distribution over each row. If so,

$$p_{mn} = p_m q_n \tag{A.8}$$

where $q_n = p(R_n)$. Since $\Sigma_n\, q_n = 1$, the two-dimensional distribution is a product of two one-dimensional distributions. In terms of the image interpretation of the probability distribution, we have a *separable* image. We say that the stochastic variables m and n are *independent*: the distribution of one variable does not depend on the other's having assumed a definite value.

From the knowledge of the 'simultaneous' or *joint* two-dimensional distribution of $[\mathsf{m},\mathsf{n}]$, more distributions associated with m and/or n may be derived. As an example, we determine the distribution of $\mathsf{k} = \mathsf{m} + \mathsf{n}$:

$$P_k = p(\mathsf{m} + \mathsf{n} = k) = \sum_{m+n=k} p_{mn} = \sum_{m} p_{m,k-m} \tag{A.9}$$

If, in particular, m and n are independent, P_k is obtained as the *convolution* of the two distributions p_m and q_n:

$$P_k = \sum_{m} p_m q_{k-m} \tag{A.10}$$

Note that the distribution of a sum of two stochastic variables does *not* equal the sum of their distributions. The latter quantity may, however, likewise be interpreted statistically (cf. for example, equation (4.21)).

EXAMPLE A.8
The *Poisson distribution* with parameter b has frequencies

$$p_m = e^{-b}\frac{b^m}{m!}, \qquad m \in [0, \infty]$$

Show that if m and n are independent Poisson variables with parameters b and c, respectively, then $\mathsf{m}+\mathsf{n}$ is Poisson distributed with parameter $b+c$. This result is of importance in photon detection (cf. Section 1.5).

We now return to our first question concerning the frequency of a certain event in a repeated experiment. For simplicity, we consider only two events H and I, with probabilities p and $q = 1 - p$. The stochastic variable X, assuming the value 1 in H and 0 in I, thus has the distribution (p, q).

In N repetitions of the experiment, the N-dimensional stochastic variable $(\mathsf{X}_1, \ldots, \mathsf{X}_N)$ occurs, where all the X_ns have the same distribution as X, and $\mathsf{Y}^{(N)} = \Sigma_{n=1}^{N} \mathsf{X}_n$ gives the *number of occurrences* of the event H. We wish to determine the distribution of $\mathsf{Y}^{(N)}$.

It is not difficult to prove directly that the answer is

$$p_n^{(N)} = \binom{N}{n} p^n q^{N-n}, \qquad n \in [0, N] \tag{A.11}$$

– the so-called *binomial distribution*, specifying the probability of obtaining n occurrences, each of probability p, in N repetitions of the individual experiment. We use this occasion to illustrate some of the methods discussed.

Since $\mathsf{Y}^{(N)} = \mathsf{Y}^{(N-1)} + \mathsf{X}_N$, equation (A.10) shows that

$$\begin{aligned} p_n^{(N)} &= p p_{n-1}^{(N-1)} + q p_n^{(N-1)} \\ p_0^{(N)} &= q^N \end{aligned}$$

EXAMPLE A.9
Show that equation (A.11) satisfies these difference equations and that it may be proved accordingly, using mathematical induction. The proof is also possible by considering $p_n^{(N)}$ to be a 'digital image' $b[N, n]$, subjected to the two-dimensional z-transformation. The calculations are left to the assiduous reader. As a hint, the z-transform is

$$B(Z, z) = \frac{p + qz}{Zz - (p + qz)}$$

It is important to realize that two very different statistical points of view are associated with digital images. One of these is that above, considering the image as a two-dimensional stochastic variable, the distribution of which gives information as to the outcome of the 'experiment' that an image photon is recorded. A specific digital image is thus, after normalization, an 'estimate' (cf. Section A.6) for the distribution. An exact knowledge of the distribution would require infinitely many digital images of exactly the same subject under exactly the same circumstances.

The other point of view, introduced in Section 1.5, regards a digital image as consisting of a large number of stochastic variables, that is, one for each pixel. Especially in the one-dimensional case, the term *stochastic process* is used; for images, the term *stochastic field* is sometimes heard. On substitution of the variable in each pixel by its *mean value* – a quantity to be defined in the next section – we return to the simple notion where a digital image is identical to an array of numbers.

A.5 Mean and variance

In connection with the question of the predictability of the outcome of an experiment, we are led to ask: What is the error if a stochastic variable is replaced by a constant (that is, a stochastic variable with a 100 per cent certain outcome)? To this end, consider first the original sample space $[1, N]$ endowed with the uniform distribution $p_n = 1/N$, and a stochastic variable X associated with it.

If one replaces the range of values x_n of X, $n \in [1, N]$, by one and the same number $\bar{x}$, the *average* seems a reasonable choice:

$$\bar{x} = \frac{1}{N} \sum_{n=1}^{N} \mathsf{X}[n] = \sum_{n=1}^{N} \mathsf{X}[n]\, p_n \tag{A.12}$$

This choice has the property that the deviations of X from this value – positive or negative – sum to zero:

$$\sum_{n=1}^{N} (\mathsf{X}[n] - \bar{x}) = \sum_{n=1}^{N} \mathsf{X}[n] - N\bar{x} = 0 \tag{A.13}$$

The average may also be evaluated by summing over the identical values x_m, $m \in [1, M]$ (cf. equation (A.4)):

$$\bar{x} = \sum_{n=1}^{N} \mathsf{X}[n]\, p_n = \sum_{m=1}^{M} x_m q_m \tag{A.14}$$

where q_m, $m \in [1, M]$, is the probability distribution defined by X.

If now $\mathsf{Z} = \mathsf{Y} \circ \mathsf{X}$ (with the same notation as previously) is composed of X followed by the stochastic variable Y, we similarly have

$$\bar{z} = \sum_{n=1}^{N} \mathsf{Z}[n]\, p_n = \sum_{n=1}^{N} \mathsf{Y}[\mathsf{X}[n]]\, p_n = \sum_{m=1}^{M} \mathsf{Y}[x_m]\, q_m \tag{A.15}$$

This equation generalizes the average to the case where the original distribution of the stochastic variable is not uniform. We therefore define the *mean* or *expectation* of a stochastic variable Y to be

$$E(\mathsf{Y}) = \sum_{m=1}^{M} \mathsf{Y}[m] q_m \tag{A.16}$$

where Y once again, for simplicity, is assumed to be defined on $[1, M]$. If Y instead is defined on $x_1, \ldots, x_M$, $\mathsf{Y}[m]$ should be replaced by $\mathsf{Y}[x_m]$, as stated in equation (A.15). Summation over the identical function values y_l, $l \in [1, L]$, leads to the alternative definition

$$E(\mathsf{Y}) = \sum_{l=1}^{L} y_l r_l \tag{A.17}$$

where r_l is the distribution of Y – or, rather, the distribution defined or induced by Y.

Further composition of stochastic variables yields nothing new. The result in question may be written as follows:

$$E(f(\mathsf{Y})) = \sum_{m=1}^{M} f(\mathsf{Y}[m]) q_m = \sum_{l=1}^{L} f(y_l) r_l \tag{A.18}$$

where f may be viewed as a function or a stochastic variable on $[1, L]$ or $(y_1, \ldots, y_L)$.

The use of the average as the basic notion for the definition of the mean is further justified by the following consideration. If, instead of discussing the deviation (equation (A.13)), one focuses upon the *signal distance* (Section 2.5) between X and the constant x, this quantity is given by

$$|\mathsf{X} - x|^2 = \sum_n (x_n - x)^2 p_n = \sum_n x_n^2 p_n - 2x \sum_n x_n p_n + x^2$$

This distance is a second-order polynomial in x, assuming its minimum value for $x = -(-2\Sigma\, x_n p_n/2) = \Sigma\, x_n p_n$, that is, for the mean $E(\mathsf{X})$. The minimum value in question,

$$V(\mathsf{X}) = \sum x_n^2 p_n - \left(\sum x_n p_n\right)^2 = E(\mathsf{X}^2) - (E(\mathsf{X}))^2 \tag{A.19}$$

is called the *variance* of X. As a measure of the difference between X and its mean, one often uses the *standard deviation* $\sigma(\mathsf{X}) = \sqrt{V(\mathsf{X})}$, that is, the signal distance between X and the constant $E(\mathsf{X})$.

For linear combinations of stochastic variables, we have

$$E\left(\sum a_n \mathsf{X}_n\right) = \sum a_n E(\mathsf{X}_n) \tag{A.20}$$

and, if the variables involved are mutually independent,

$$V\left(\sum a_n X_n\right) = \sum a_n^2 V(X_n) \tag{A.21}$$

The latter result is obtained by (repeated) application of equation (A.10).

EXAMPLE A.10

Given the probability distribution $p_1 = 0.3$, $p_2 = p_3 = 0.2$, $p_4 = p_5 = p_6 = 0.1$ on the interval $[1, 6]$, together with the two stochastic variables X and Y, find the mean value and variance for the stochastic variables X + Y in the following two cases:

(a) $X[n] = n$ and $Y[n] = n^2$ (where X and Y are assumed to be independent)
(b) $X[n] = n$ and $Y = X^2$.

EXAMPLE A.11

Given n identically distributed and independent stochastic variables $X_1, \ldots, X_N$, the mean and variance of which are μ and σ^2, respectively, their *average* $\bar{X} = (X_1 + \cdots + X_N)/N$ is also a stochastic variable. Show that its mean and variance are given by

$$E(\bar{X}) = \mu \qquad \text{and} \qquad V(\bar{X}) = \sigma^2/N \tag{A.22}$$

The average thus has the same mean as the individual variables; the standard deviation, however, is decreased by a factor of $\sqrt{N}$.

Note that even if $X_1, \ldots, X_N$ are identically distributed, they should be considered as describing separate experiments. As regards notation, the situation is highlighted by insisting that $X + X + \cdots + X$ (N terms) is *not* equal to NX. Both have mean value $N\mu$, but the former has the variance $N\sigma^2$, the latter $N^2\sigma^2$.

A.6 Moments

The point of view adopted when introducing the concepts of mean and variance was the characterization of stochastic variables with respect to their degree of 'concentration' around a certain number. Since, in the present context, a probability distribution is defined by a stochastic variable, this characterization must apply to the distribution as such, irrespective of the stochastic variable generating it.

As mentioned above, the concepts 'probability distribution' and 'distribution of the stochastic variable n given by $n[n] = n$' are identical. Consequently, we define the mean of the distribution $p = (p_1, \ldots, p_N)$ by

$$E(p) = E(\mathsf{n}) = \sum_{n=1}^{N} np_n \tag{A.23}$$

and its variance by

$$V(p) = V(\mathsf{n}) = \sum_{n=1}^{N} n^2 p_n - \left(\sum_{n=1}^{N} np_n\right)^2 \tag{A.24}$$

that is, as $V(p) = E(\mathsf{n}^2) - E(\mathsf{n})^2$.

EXAMPLE A.12

The mean and variance of the uniform distribution on $[1, N]$ are given by

$$E(\mathsf{n}) = \sum_{n=1}^{N} n \cdot \frac{1}{N} = \frac{(N+1)}{2}$$

$$E(\mathsf{n}^2) = \sum_{n=1}^{N} n^2 \cdot \frac{1}{N} = \tfrac{1}{6}(N+1)(2N+1)$$

$$V(\mathsf{n}) = \tfrac{1}{6}(N+1)(2N+1) - \tfrac{1}{4}(N+1)^2 = \frac{N^2-1}{12}$$

EXAMPLE A.13

Find the mean and variance of the so-called *geometric distribution*

$$p_n = (1-a)a^n, \qquad n \in [0, \infty]$$

(where $|a| < 1$), in the case $a = \frac{1}{3}$.

EXAMPLE A.14

The mean and variance of the binomial distribution are

$$E(\mathsf{n}) = Np \qquad \text{and} \qquad V(\mathsf{n}) = Npq$$

Show this, first directly from equation (A.11) and the above formulae, next by evaluating the mean and variance for the distribution $(p_1, p_2) = (p, q)$ and using the remark at the end of Section A.5.

The mean and variance for the quantity n/N, used for the very definition of the probability concept, are thus, respectively, p (as it should be) and pq/N (converging towards zero).

The quantities $E(\mathsf{n})$ and $E(\mathsf{n}^2)$ are examples of distribution *moments*. The moment P_k of *order* k for the distribution p is defined as

$$P_k = E(\mathsf{n}^k) = \sum_{n=1}^{N} n^k p_n \tag{A.25}$$

and obviously $P_0 = 1$, $P_1 = E(p)$ and $P_2 - P_1^2 = V(p)$.

The moments are frequently evaluated from a shifted zero point x, so that in lieu of (A.26) one should use $E((\mathsf{n} - x)^k) = \Sigma (n - x)^k p_n$. If, in particular, x is the mean $\mu = E(p)$, we speak of the *central* moments of the distribution:

$$P_k^c = E((\mathsf{n} - \mu)^k) = \sum_{n=1}^{N} (n - \mu)^k p_n \tag{A.26}$$

Here, $P_1^c = 0$ and $P_2^c = V(p)$, and of course $P_0^c = 1$.

The significance of the moments is to be found in the fact that they provide a complete characterization of the distribution (see Section A.8).

EXAMPLE A.15

Express P_3^c in terms of P_1, P_2, and P_3.

For probability distributions of dimension 2 or higher, the moments assume correspondingly higher dimension (number of indices). A two-dimensional distribution p_{mn} has, for example, moments

$$P_{kl} = \sum_{m=1}^{M} \sum_{n=1}^{N} m^k n^l p_{mn} \tag{A.27}$$

but if it is generated as a simultaneous distribution for two independent variables, that is, $p_{mn} = p_m q_n$, P_{kl} splits into the product

$$P_{kl} = P_k Q_l \tag{A.28}$$

of moments of order k and l, respectively, for the individual distributions.

The most commonly occurring two-dimensional moments for a distribution p_{mn} are its two means

$$P_{10} = \sum_m \sum_n m p_{mn} \qquad \text{and} \qquad P_{01} = \sum_m \sum_n n p_{mn} \tag{A.29}$$

which together define the *centre* (P_{10}, P_{01}) of the distribution, and its *covariance*

$$\begin{aligned} P_{11}^c &= \sum_{mn} mn p_{mn} - \left(\sum_{mn} m p_{mn}\right)\left(\sum_{mn} n p_{mn}\right) \\ &= P_{11} - P_{10} P_{01} \end{aligned} \tag{A.30}$$

EXAMPLE A.16

Show that if m and n are independent stochastic variables, their covariance is $C(\mathsf{m},\mathsf{n}) = E(\mathsf{mn}) - E(\mathsf{m})E(\mathsf{n}) = 0$.

EXAMPLE A.17

Find the covariance for the two variables m and $\mathsf{n} = \mathsf{m}^2$, where m has a uniform distribution in $[0, 3]$.

A.7 Tests and estimation

In a *test*, a *hypothesis* is examined by means of statistical experiments. The hypothesis concerns the distribution of one or more stochastic variables, and should be formulated based upon a presumed real cause/effect relationship, *never* from 'statistical material'. A properly formulated test is always accompanied by an *alternative hypothesis*, which should be chosen so as to be of maximum effect in disproving the hypothesis itself – if corroborated by the experiments.

The method consists in choosing an appropriate stochastic variable, a so-called *statistic*. This statistic T should be chosen so as to cause a large deviation between its distributions $p^{(H)}$ and $p^{(A)}$. Here, $p^{(H)}$ ($p^{(A)}$) is the correct distribution provided the hypothesis H is true (the alternative A is true), and the ideal test conditions correspond to both $V(p^{(H)})$ and $V(p^{(A)})$ being much less than $(E(p^{(H)}) - E(p^{(A)}))^2$.

In the traditional procedure, the hypothesis is assumed true 'until the contrary has been proved'. This means that, before the test, a small probability s is chosen, the so-called *significance limit*; the canonical value is $s = 5\%$. With this limit, we evaluate the interval $[t_-, t_+]$ of t-values $t_1, t_2, \ldots$ for which $p^{(H)}([t_-, t_+]) \leqslant 0.05$ and $p^{(A)}([t_-, t_+])$ is maximized. This interval is called the *critical range* of the test. If T is observed to fall within it, the hypothesis H is rejected.

EXAMPLE A.18

I have inherited a die from a gambler, and I suspect it to be biased in such a way that it yields a 5 in half of the trials and any other number in 1/10 of the trials. I would like to test this hypothesis statistically.

As the *alternative*, I choose the hypothesis that the die is unbiased, that is, every number occurs with probability 1/6. The statistic I choose is the number n of 5s in N throws, and the significance limit is taken to be 5%.

Under the hypothesis H, n is binomially distributed with parameter $p = 1/2$ (hence also $q = 1/2$); under the alternative A, n is binomially distributed with $(p, q) = (1/6, 5/6)$:

$$p_n^{(H)} = \binom{N}{n} \left(\tfrac{1}{2}\right)^N$$

$$p_n^{(A)} = \binom{N}{n} \left(\tfrac{1}{6}\right)^n \left(\tfrac{5}{6}\right)^{N-n}$$

These two alternatives are tabulated below for $N = 5$.

n	0	1	2	3	4	5
$p_n^{(H)}$	0.0313	0.1563	0.3125	0.3125	0.1563	0.0313
$p_n^{(A)}$	0.4019	0.4019	0.1608	0.0322	0.0032	0.0001

Sketch the two distributions. The critical range, where the hypothesis H is rejected, consists of the outcome $n = 0$ only. (Why shouldn't $n = 5$ be included?)

There is, unfortunately, only time for 5 throws; I obtain 3, 5, 6, 3, and 2, that is, $n = 1$. From this material, I can only conclude that the hypothesis should not be rejected.

One should realize that this procedure only makes sense if the hypothesis of a biased die is substantiated by physical or other evidence. Given only the very sparse experimental material above, it would be absurd to conclude that the die is biased. If it is *known* to be biased, the material points to the value 3 as the most probable number of spots (an example of an *estimation* of a distribution parameter, see below).

Clearly, it is not difficult – by manipulating significance limits, hypotheses with and without alternatives, and the size of the corpus of data – to accept or reject anything. The literature has witnessed a horrifying number of 'scientific' statistical investigations and a host of meaningless conclusions based upon superficial knowledge of the principles behind statistical testing of hypotheses.

For concepts such as *optimal tests* and type I and type II errors, the reader is referred to p. 135 and to Sections 4.5.1 and 5.1.

In an *estimation*, one attempts to characterize the distribution of a stochastic variable X. The available information is limited to a set of observations of it, $x_1, \ldots, x_N$. The characterization should consist in the specification of one or more *parameters* in the distribution under consideration; normally, one does not question the analytical form of the distribution (a Poisson distribution, a binomial distribution, or whatever). If, for example, n is known to be Poisson distributed, we naturally wish to 'estimate' the distribution parameter. To this end, an *estimator* is employed. An estimator for a parameter a in the distribution of a stochastic variable X is itself a stochastic variable dependent on the one given:

$$\mathsf{Y} = \mathsf{Y}(\mathsf{X})$$

for which, ideally, $E(\mathsf{Y}) = a$ and $V(\mathsf{Y})$ is minimized.

EXAMPLE A.19

In a certain pixel in a digital image, the number of photons n is detected – an observation of the Poisson variable n. If the distribution parameter is b, n is always used in practice as an estimator for b, if no other information is available. Naturally, $E(\mathsf{n}) = b$. The variance $V(\mathsf{n})$ also equals b.

If, however, several pixels with distribution parameter b exist, one may form the average of the corresponding observations $n_1, \ldots, n_N$. This average, $\bar{n} = (n_1 + \cdots + n_N)/N$, is an observation of $\bar{\mathsf{n}} = (\mathsf{n}_1 + \cdots + \mathsf{n}_N)/N$. This stochastic variable, likewise, has mean

$$E(\bar{\mathsf{n}}) = \frac{1}{N} \sum_{n=1}^{N} b = b$$

but as the variance is

$$V(\bar{\mathsf{n}}) = \frac{1}{N^2} \sum_{n=1}^{N} b = b/N$$

it is a more useful estimator than the single-pixel observation. Note that $\bar{\mathsf{n}}$ is no longer Poisson distributed.

A commonly used estimation method employs the *maximum likelihood* (ML) principle. For a given distribution parameter a, the observation x of X has a definite probability $\mathscr{L}_x$. But if a is envisaged as varying, this probability also varies:

$$\mathscr{L}_x = \mathscr{L}_x(a)$$

This function of a, which depends on the actual distribution and on x, is called the *likelihood function* for a.

The maximum likelihood estimator $\hat{a}$ for a is the value for which the likelihood function is maximized. A necessary condition is

$$\mathscr{L}_x'(\hat{a}) = 0 \tag{A.31}$$

EXAMPLE A.20

Given K observations $n_1, \ldots, n_K$ of a Poisson variable n, if its distribution parameter is b, the probability p_1 of obtaining the observation n_1 is

$$p_1 = \mathrm{e}^{-b} \frac{b^{n_1}}{n_1!}$$

and similarly for $n_2, \ldots, n_K$. The total probability for the entire set is

$$\mathscr{L}(b) = \prod_{k=1}^{K} \mathrm{e}^{-b} \frac{b^{n_k}}{n_k!}$$

which is to be maximized. Instead, we maximize $L(b) = \ln(\mathscr{L}(b))$, with the same result, since ln is an increasing function:

$$L(b) = \sum_{k=1}^{K} (-b + n_k \ln b) - \sum \ln(n_k!)$$

so that

$$L'(b) = -K + \frac{1}{b} \sum_{k=1}^{K} n_k$$

This quality is 0 for $b = (\Sigma_1^K n_k)/K$, where the likelihood function is a maximum.

The stochastic variable $\bar{\mathsf{n}} = (\Sigma_1^K \mathsf{n}_k)/K$ is thus the maximum likelihood estimator for b.

EXAMPLE A.21

The stochastic variable n is known to obey a geometric distribution (see Example A.13). It is observed in five independent experiments, and the results are

$$3, 2, 7, 1, 7$$

Find the distribution parameter a by ML estimation.

A.8 Continuous probability distributions

The results derived so far apply, with minor changes, to the limiting case where the sample spaces are intervals of rational or real numbers.

If a stochastic variable X assumes values in the interval $(0, T)$, this interval is divided into N equal parts, resulting in a probability distribution

$$p_n = p((n-1)\,T/N \leqslant \mathsf{X} \leqslant nT/N)$$

p_n being the probability of X falling in the nth subinterval. We now define a function f_N by

$$f_N(x) = \frac{N}{T} p_n, \qquad x \in ((n-1)T/N, nT/N) \tag{A.32}$$

and the limit $N \to \infty$ establishes a function f on the interval $(0, T)$. From expression (2.7) we obtain

$$\frac{1}{N} \sum_{n=1}^{N} f(nT/N) \to \frac{1}{T} \int_0^T f(t)\, \mathrm{d}t$$

But as

$$\frac{T}{N} \sum_{n=1}^{N} f(nT/N) = \sum_{n=1}^{N} p_n = 1$$

we must have

$$\int_0^T f(t)\, \mathrm{d}t = 1 \tag{A.33}$$

The function f thus defined is called the *probability density* for X. Its integral

$$F(x) = \int_0^x f(t)\, \mathrm{d}t \tag{A.34}$$

is called the *(cumulative) distribution function* for X; it is non-decreasing, positive and obeys $F(0) = 0$, $F(T) = 1$. The same notation is used if the probability distribution is discrete (digital!) and has frequencies p_n, $n \in [1, N]$:

$$P[n] = \sum_{k=1}^{n} p_k \tag{A.35}$$

If now X is a stochastic variable with the probability density $f(t)$, $t \in (0, T)$, and I is a subinterval $(t_1, t_2) \subseteq (0, T)$, then

$$p(\mathsf{X} \in I) = \int_{t_1}^{t_2} f(t)\, \mathrm{d}t = F(t_2) - F(t_1) \tag{A.36}$$

where the left-hand side is the probability of X falling within I, that is, between t_1 and t_2. In particular, we may interpret the density as the probability of X falling between t and $t + \mathrm{d}t$:

$$p(t \leqslant \mathsf{X} \leqslant t + \mathrm{d}t) = f(t)\, \mathrm{d}t \tag{A.37}$$

An equally useful result is

$$p(\mathsf{X} \leqslant x) = F(x) \tag{A.38}$$

This relation may be used for determining the distribution of a composite stochastic variable $Y = h(X)$. If X and Y have distribution functions F and G, respectively, the following holds

$$G(y) = p(Y \leqslant y) = p(h(X) \leqslant y) \\ = p(X \leqslant h^{-1}(y))$$

where it has been assumed that h is a monotonic function. This means that

$$G(y) = F(h^{-1}(y)) \tag{A.39}$$

This rule for the transformation of stochastic variables is most conveniently expressed in terms of probability densities. Differentiation of equation (A.39) with respect to y gives

$$G'(y) = F'(h^{-1}(y)) \frac{dh^{-1}(y)}{dy}$$

where $G' = g$ and $F' = f$ are the densities for Y and X, respectively. With $y = h(x)$ (or $x = h^{-1}(y)$), this formula shows that

$$g(y) = f(x) \frac{dx}{dy} \tag{A.40}$$

which is the golden rule for transformation of probability densities.

The formula (A.41) is easier to remember in the symbolic form

$$g(y)\, dy = f(x)\, dx \tag{A.41}$$

which, in view of equation (A.37), possesses an obvious intuitive interpretation.

EXAMPLE A.22

The stochastic variable X is uniformly distributed in the interval $(-\frac{1}{2}\pi, \frac{1}{2}\pi)$. Find the distribution of $Y = \sin X$.

Solution. Formula (A.41) gives

$$g(y) = \frac{1}{\pi} \frac{dx}{dy} = \frac{d(\arcsin y)}{dy}$$

$$= \frac{1}{\pi} \frac{1}{\sqrt{1 - y^2}}, \qquad y \in (-1, 1)$$

The other definitions, concepts and results may immediately be taken over by the continuous distributions, if probabilities are replaced by densities, sums by integrals, and so on. For instance, the criterion for the *independence* of two stochastic variables X_1 and X_2 is that their 'simultaneous density' $f(x_1, x_2)$ should be a product of the individual densities:

$$f(x_1, x_2) = f_1(x_1)f_2(x_2) \tag{A.42}$$

Moments, in particular the mean, are defined in terms of integrals instead of sums:

$$F_k = \int t^k f(t)\, dt$$

Finally, it is mentioned in the context of ML estimation that the likelihood function for a parameter a, corresponding to the observations $x = (x_1, \ldots, x_N)$, is given by $\mathscr{L}_x = f(x_1)f(x_2)\ldots f(x_N)$ (where the explicit dependence of a has been omitted).

EXAMPLE A.23

Find the mean and variance for the distributions having densities (a) $f(t) = 1$, $t \in (0, 1)$, and (b) $f(t) = a\mathrm{e}^{-at}$, $t \in (0, \infty)$.

We conclude this section by arguing that a distribution is completely determined by its moments. To this end, consider the Fourier transform φ for the probability density f:

$$\varphi(\omega) = \int f(t)\mathrm{e}^{-\mathrm{i}\omega t}\, dt$$

(The symbol φ has been chosen in order to include the factor $\sqrt{2\pi}$ and to avoid confusion with the distribution function and the moments F_k.) Differentiation k times yields

$$\varphi^{(k)}(\omega) = \int f(t)(-\mathrm{i}t)^k \mathrm{e}^{-\mathrm{i}\omega t}\, dt = (-\mathrm{i})^k \int t^k f(t)\mathrm{e}^{-\mathrm{i}\omega t}\, dt$$

in particular,

$$\varphi^{(k)}(0) = (-\mathrm{i})^k F_k$$

According to Taylor's formula, $\varphi(\omega)$ is completely specified by the derivatives, evaluated at 0:

$$\begin{aligned}\varphi(\omega) &= \sum_{k=0}^{\infty} \varphi^{(k)}(0)\omega^k/k! \\ &= \sum_{k=0}^{\infty} (-\mathrm{i}\omega)^k F_k/k! \end{aligned} \tag{A.43}$$

whence f, to be obtained by inverse Fourier transformation, is uniquely characterized by means of its moments.

A.9 The normal distribution

One of the most fundamental results in statistics is the *central limit theorem*. This theorem says that, on summation of 'several' stochastic variables with the same distribution, one obtains a new one, the distribution of which is independent of the original ones. This distribution is called the *normal distribution* (or the Gaussian[1] distribution) and has density

$$f(x) = \frac{1}{\sqrt{2\pi}\sigma} e^{-(x-\mu)^2/2\sigma^2}, \qquad x \in (-\infty, \infty) \tag{A.44}$$

One can easily show that the two parameters μ and σ^2, characterizing the distribution, equal the mean and variance, respectively. A little more formally stated, the central limit theorem says that if $\mathsf{X}_1, \mathsf{X}_2, \ldots$ are identically distributed stochastic variables, the distribution of the sum

$$\mathsf{Y}^{(N)} = \sum_{n=1}^{N} \mathsf{X}_n$$

will approach a normal distribution as N tends to ∞. Remarkably, the sum converges towards a normal distribution regardless of the form of the original distribution.

EXAMPLE A.24

The binomial distribution (equation (A.11)) has been produced by addition of identically distributed variables. Thus, for large N, it may be approximated by the normal distribution. Since the mean and variance for the binomial distribution is given by Example A.14, the normal distribution in question must satisfy $\mu = Np$ and $\sigma^2 = Npq$.

We next show that the Poisson distribution can be approximated by a binomial distribution – and consequently by a normal distribution. If the Poisson distribution parameter is b, and it is to be approximated by a binomial distribution of order N, we must choose

$$p = \frac{b}{N}$$

in order to comply with $\mu = Np = b$. Next, $\sigma^2 = Npq = Np(1 - p) = b$ also holds for

[1] C. F. Gauss, German mathematician, 1777–1855.

large N. The ratio between the two distributions is now

$$\frac{p_n^{(\text{binomial})}}{p_n^{(\text{Poisson})}} = \frac{\binom{N}{n}(b/N)^n(1-b/N)^{N-n}}{e^{-b}b^n/n!}$$

$$= \frac{N(N-1)\dots(N-n+1)}{N^n} \frac{1}{(1-b/N)^n} \frac{(1-b/N)^N}{e^{-b}}$$

where each of the three factors converges towards 1 as $N \to \infty$, if n is fixed.

EXAMPLE A.25

Show that $(1-b/N)^N \to e^{-b}$ as $N \to \infty$.

Hint: Use the fact that $\ln(1+x)/x \to 1$ for $x \to 0$.

From this it follows that the Poisson distribution may be approximated by

$$p_n \approx \frac{1}{\sqrt{2\pi b}} e^{-(n-b)^2/2b} \tag{A.45}$$

as stated in Table 5.1.

Finally, we exemplify some of the basic statistical concepts by means of the normal distribution.

The *sum* of two normally distributed stochastic variables is again normally distributed. If the variables in question are denoted X_1 and X_2, their parameters (μ_1, σ_1^2) and (μ_2, σ_2^2), and their densities f_1 and f_2, respectively, the density for $X = X_1 + X_2$ becomes

$$f(x) = \int f_1(x-y) f_2(y)\, dy$$

$$= \frac{1}{\sqrt{2\pi}\sigma_1} \frac{1}{\sqrt{2\pi}\sigma_2} \int e^{-(x-y)^2/2\sigma_1^2} e^{-y^2/2\sigma_2^2}\, dy$$

$$= \frac{1}{\sqrt{2\pi}\sigma} e^{-(x-\mu)^2/2\sigma^2} \tag{A.46}$$

with the parameters

$$\mu = \mu_1 + \mu_2 \qquad \text{and} \qquad \sigma^2 = \sigma_1^2 + \sigma_2^2 \tag{A.47}$$

EXAMPLE A.26

Fill in the details in the above calculation.

The *moments* of the normal distribution are determined from the moments of the *standard normal distribution*, the density of which is

$$f(x) = \frac{1}{\sqrt{2\pi}} \mathrm{e}^{-x^2/2} \tag{A.48}$$

corresponding to parameters $(\mu, \sigma^2) = (0, 1)$. Its moments are

$$f_n = \begin{cases} 0 & n \text{ odd} \\ (n-1)(n-3)\dots 3\cdot 1 & n \text{ even} \end{cases} \tag{A.49}$$

EXAMPLE A.27
Show this, for example by differentiation of

$$s^{-1/2} = \frac{1}{\sqrt{\pi}} \int_{-\infty}^{\infty} \mathrm{e}^{-st^2}\, \mathrm{d}t$$

with respect to s.

By way of an example of ML estimation, we find the standard deviation σ for a normally distributed stochastic variable, with $\mu = 0$, from N independent observations $x_1, \dots, x_N$ of it. The likelihood function then becomes

$$\mathscr{L}(\sigma) = \prod_{n=1}^{N} f(x_n) = \frac{1}{(\sqrt{2\pi})^N \sigma^N} \mathrm{e}^{-\sum x_n^2/2\sigma^2}$$

and the log-likelihood function

$$L = \ln(\mathscr{L}) = -\frac{N}{2}\ln(2\pi) - N\ln\sigma - \sum x_n^2/2\sigma^2$$

The ML estimator $\hat{\sigma}$ for σ is thus found from

$$L'(\hat{\sigma}) = -\frac{N}{\hat{\sigma}} + \frac{1}{\hat{\sigma}^3}\sum x_n^2 = 0$$

with solution

$$\hat{\sigma}^2 = \frac{1}{N}\sum x_n^2 \tag{A.50}$$

There exists an extensive body of theory for *tests* based upon the normal distribution and its derived distributions. To conclude this section, Table A.1 gives a few important values for the standard normal distribution $f(x)$ (equation (A.48)) and its distribution function $F(x)$.

Table A.1 The normal distribution.

x	0	1	1.282	1.645	2	2.326	3
$f(x)$	0.3989	0.2420	0.1754	0.1031	0.0540	0.0267	0.0044
$F(x)$	0.5000	0.8413	0.9000	0.9500	0.9772	0.9900	0.9987

EXAMPLE A.28

A statistic T is normally distributed with parameters $(\mu, \sigma^2) = (2, 9)$, under the hypothesis H. Under the alternative A, the distribution is the standard normal distribution. A significance limit of 5% is chosen. What is the critical range for this test?

APPENDIX B

Programs

B.1 A fast Fourier transform algorithm

```
SUB FTT(fr(1),fi(1),ln)
INTEGER i-n
pi=4*ATN(1)
n=2^ln
nv2=n/2
nm1=n-1
j=1
FOR i=1 TO nm1                  (a)
IF (i>=j) THEN GOTO 1
SWAP fr(i),fr(j)
SWAP fi(i),fi(j)
1 : k=nv2
2 : IF (k>=j) GOTO 3
j=j-k
k=k/2
GOTO 2
3 : j=j+k
NEXT i
FOR l=1 TO ln                   (b)
le=2^l
le1=le/2
ur=1
ui=0
wr=COS(pi/le1)
wi=-SIN(pi/le1)                 (c)
FOR j=1 TO le1
FOR i=j TO n STEP le
ip=i+le1
tr=fr(ip)*ur-fi(ip)*ui
ti=fr(ip)*ui+fi(ip)*ur
fr(ip)=fr(i)-tr
fi(ip)=fi(i)-ti
fr(i)=fr(i)+tr
fi(i)=fi(i)+ti
NEXT i
ur1=ur*wr-ui*wi
ui=ur*wi+ui*wr
ur=ur1
NEXT j
NEXT l
div=1/SQR(n)                    (d)
FOR i=1 TO n
fr(i)=fr(i)*div
fi(i)=fi(i)*div
NEXT i
END SUB
```

COMMENTS

The subprogram replaces the complex signal $f(1), f(2), \ldots, f(n)$, in which $n = 2^{ln}$, by its digital Fourier transform. The signal f is given by real part `fr` and imaginary part `fi`.

(a) Swapping of the original data, to be used in
(b) a calculation of the DFT by successively halving its order.
(c) For the inverse DFT, **-SIN** should be replaced by **SIN**.
(d) Division by $\sqrt{n}$.

B.2 Information content and entropy

```
SUB info (n,p(1),H,J)
psum=0 : hsum=0 : jsum=0
FOR m=1 TO n
psum=psum+p(m)
hsum=hsum+p(m)*LOG(p(m))                              (a)
NEXT m
H=(LOG(psum)-hsum/psum)/LOG(2)
FOR m=n TO 2 STEP -1
FOR l=1 TO m-1
IF p(l)<p(l+1) THEN SWAP p(l),p(l+1)                  (b)
NEXT l
FOR l=1 TO m-2
IF p(l)<p(l+1) THEN SWAP p(l),p(l+1)                  (b)
NEXT l
p(m-1)=p(m-1)+p(m)                                    (c)
jsum=jsum+p(m-1)                                      (d)
NEXT m
J=jsum/psum
END SUB
```

COMMENTS

The subprogram calculates entropy $\mathscr{H}$ and information content $\mathscr{J}$ for the distribution $p(1), p(2), \ldots, p(n)$. The ps are assumed to be *positive*, but not to be normalized.

(a) Calculation of $\mathscr{H}$ according to definition.
(b) Sorting of the ps according to value; the two smallest are assigned the highest indices.
(c) Calculation of a new probability distribution according to the Huffman principle.
(d) Summation of code word lengths.

The above calculation of $\mathscr{J}$ has been chosen for clarity and is not optimal with respect to computing time.

APPENDIX C

Solutions to Examples

Chapter 1

1.1

Field area	1/49	9/49	25/49	1	9	25	49
Grain density	0	16.3	15.7	8	6.4	7.1	7.2

1.2 14 and 47.

1.4 4.

1.6 6×10^{-9} W m^{-2}.

1.10 20 dB.

1.11 Approx. 8000; 56 s.

1.12 The probability of detecting two or more photons is of second order in β.

1.13 $\beta = 0.43$. (a) 0.90 (b) 0.42 (c) 0.013.

Chapter 2

2.3 The lower left-hand corner coordinates are $(-22, 4)$.

2.5 (a) and (c).

2.15 40 samples.

2.16

Sample	$f[0]$	$f[1]$	$f[10]$	$f[20]$	$f[39]$
End point	0.00	1.09	5.00	0.00	-1.09
Mid-point	0.55	1.62	4.83	-0.24	-0.55
Average	0.55	1.62	4.82	-0.24	-0.55

2.17 $3\delta[n-2] - 2\delta[n+1] = 3s[n-2] - 3s[n-3] - 2s[n+1] + 2s[n]$.

2.18 Both j and k are parameters.
2.20 $\bar{b}(x, y) = (11r_1(x+2) + 26r_1(x+1) + 18r_1(x) + 5r_1(x-1)) \cdot (15r_1(y+1.5) + 9r_1(y+0.5) + 13r_1(y-0.5))$.
2.22 $|f| = 4$, $|g| = \sqrt{3}$, $|g - f| = \sqrt{23}$, $f \cdot g = g \cdot f = -2$.
2.23 $\sqrt{6}/2$.
2.24 (a) 4.40 and 0.63, (b) 2.77 and 0.40.
2.25 (a) 0.63 (b) 0.63 (c) 0.63 (exponential signal);
(a) 0.37 (b) 0.39 (c) 0.40 (harmonic signal).
2.27 Analog: $|b| = 15.2$, $|c| = 22.8$, $b \cdot c = 0$, $|c - b| = 27.4$.
Mid-point sampling: $|b| = 13.4$, $|c| = 21.0$, $b \cdot c = 0$, $|c - b| = 24.9$.
Averaging: $|b| = 13.9$, $|c| = 21.4$, $b \cdot c = 0$, $|c - b| = 25.5$.
2.30 $(-2, 2, 8)$.
2.32 $a_0 = 0.05$, $a_1 = -0.6$, $a_2 = 1.5$.
2.34 The coefficients a_- and a_+ are, in the analog case, -0.03 and 0.87.
In the three digital cases: (a) -0.05 and 0.82, (b) -0.03 and 0.87, (c) -0.03 and 0.88.
2.37 The coefficients are, in all four cases, 0, $-\mathrm{i}/2$, 0 and $\mathrm{i}/2$.
The errors are (a) $\sqrt{2}$, (b) $\sqrt{3}/2$, (c) $\sqrt{3}/2$ and (d) 0.

2.39 $$|t| = \tfrac{1}{2} - \frac{4}{\pi^2} \sum_{n=0}^{\infty} \frac{\cos(2n+1)\pi t}{(2n+1)^2}, \qquad t \in (-1, 1).$$

2.41 $f(2.5) = -3.1$ and $f(7.5) = -3.8$.

Chapter 3

3.1 (b) and (c).
3.2 (a) 2, (b) 26, (c) 17.
3.3

n	-1	0	1
$UTf[n]$	-9	0	1
$TUf[n]$	-18	-3	0

3.8 $\frac{1}{7}(6, -5, 1)$, $\frac{1}{7}(1, 5, -1)$ and $\frac{1}{7}(-1, 2, 1)$.
3.9 $(2, 4\mathrm{i}, 0, -2\mathrm{i})$.
3.12

(a) $\frac{1}{3}$

$-w$	2	$-2(1+w)$
$1+w$	$2w$	2
5	$1+w$	$-w$

$(w = w_3 = \mathrm{e}^{2\pi \mathrm{i}/3})$

(b) $\frac{1}{3}$

$-2w$	4	$-4(1+w)$
$2(1+w)$	$4w$	4
1	$2(1+w)$	$-2w$

3.15 $F(\omega) = \sqrt{\frac{\pi}{2}}\, e^{-|\omega|}$.

3.18 $512 \times 5120 + 1024 \times 2304 =$ approx. 5 million.
With separability: $5120 + 2304 + 512 \times 1024 =$ approx. 500 000.

3.22 (a) $g[n] = \delta[n+1] + 4\delta[n] + 6\delta[n-1] + 4\delta[n-2] + \delta[n-3]$,
(b) $g[n] = \delta[n+3] + 3\delta[n+2] + 4\delta[n+1] + 4\delta[n] + 3\delta[n-1] + \delta[n-2]$,
(c) $g[0] = 3,\ g[1] = 4,\ g[2] = 4$,
(d) $g[0] = 9/2,\ g[1] = 9,\ g[2] = 18$.

3.24 Pixel within: 262.2; outside: 140 (before); 142.2 (after).

3.25 (a) $H(z) = z^{-1} + 2 + z$, (b) $H(z) = 3z/(3z-1)$.

3.27 (a) $B(u, v) = 3uv/(3u-1)(v-1)$, (b) $B(u, v) = uv^5/(uv^2-1)$.

3.28 (a) $f[n] = \delta[n-1] + 2ns[n]$, (b) $f[n] = 3\delta[n+1] + (-1)^n(2-3^n)s[n]$.

3.30 $N(N^4-1)/30$.

Chapter 4

4.1 For instance, one could use linear regression or another type of interpolation in the Fourier coefficients.

4.3 Number of decodings: 12. Decoded sequence: MINK.

4.4 Only code 1.

4.7 (a) 0.47, (b) 1.92, (c) 1.58, and (d) 1.75.

4.11 $\mathcal{J} = 2$, $\mathcal{H} = 1.955$.

Probability	0.40	0.32	0.16	0.08	0.04
Code (e.g.)	0	11	101	1001	1000

4.12 (a) 3 million, (b) 2.2 million.

4.13 Minimum number of bits: approx. 1.5 million; bandwidth: 46 MHz.

4.14 (a) 1 2 3456 (1.70 bits), (b) 123 45 6 (0.93 bits).

4.15 (a) 1.03, (b) 0.56 (per pixel).

4.16

Quantization	$\mathcal{H}$	$\mathcal{J}$	Image error
original	3.089	3.109	2.236
(a)	1.946	2	0.638
(b)	1.548	1.558	0.816

4.19 All values are multiplied by i.

Chapter 5

5.2 $p = 0.85$.
5.3 2467, 10 090 and 10 522.
5.4 $p = 0.13\%$.
5.6 Yes (the 2σ-limit for the average is 10 017.9).
5.7 $b_+ = 1089.9$.
5.9 63.4%; 1-pixel probability: 0.01%.
5.12 The class defined by e_2.
5.14

1	0	1
0	4	0
1	0	1

as well as e_1' and e_2'.

5.15 The last two masks should be replaced by

0	−1	1
1	0	−1
−1	1	0

and

1	−1	0
−1	0	1
0	1	−1

5.19

(a)

$\frac{1}{6}$	$\frac{1}{3}$	$\frac{1}{6}$	$\frac{1}{3}$
$\frac{1}{6}$	$\frac{1}{3}$	$\frac{1}{6}$	$\frac{1}{3}$
$\frac{1}{2}$	0	$\frac{1}{2}$	0

, (b) $\frac{1}{2}$

1	0	1	0
0	1	0	1
1	0	1	0

5.20 (a) $c(x, y) = -(x^2 + y^2)/6 - xy/2 - |x| - |y| + 6$
(b) $c[m, n] = (55 - m(m+6))(36 - n(n+5))/36$.
5.21

Object	b_{20}	b_{11}	b_{02}	b_{20}^c	b_{11}^c	b_{02}^c
Rectangle	7/3	−1	13/12	4/3	0	1/12
Circle	145/16	−6	65/16	1/16	0	1/16

5.23 $b_{20}' = 9.61$, $b_{02}' = 2.25$.
5.24 $b_{20}^c = 0.4$, $b_{11}^c = 0.54$, $b_{02}^c = 1.02$.
5.25 Pear.
5.26 (a) 8226 ± 280, (b) 6418 ± 199.
5.27 (a) 913 ± 30, (b) 924.5 ± 21 (± 30 without the reference object).

Chapter 6

6.1 $b(1, 0) = 5.35$.
6.3 $a = L(b)$.

6.4 $a_0 = (\frac{1}{2}, -\frac{1}{2})$, mirror image: $(1, 1)$;
$a_1 = 1,\ a_{11} = a_{22} = 0,\ a_{12} = 1,\ a_2 = -1,\ a_{21} = 1.$

6.5 (a) A line symmetry about the line $(x, y) = (5, 0) + t(0, 1)$ followed by a translation by the vector $(0, 2)$.
(b) A line symmetry about the line $(x, y) = (0, 3) + t(3, -1)$ followed by a multiplication by 2, centred on $(0, 3)$.

6.6 The feature is *Prague*, located at $(-85, 708)$.

6.7 $f(t) = \frac{1}{6}(5x^3 + 3x^2 - 14x)$; $f(\frac{1}{2}) = -0.94$.

6.8 $f(t) = 2\ \mathrm{sinc}(t + \pi) - \mathrm{sinc}(t - \pi) + 4\ \mathrm{sinc}(t - 2\pi)$; $f(\frac{1}{2}\pi) = -1.91$.

6.10 40.34.

6.11 Point intensity: 27.6; pixel intensity: 28.3.

Chapter 7

7.2

p_n	0.15	0.08	0.20	0.07	0.35	0.15
p'_n	0.23		0.27		0.35	0.15

7.3 $a = 0.066$, $p'_4 = 0.14$.

7.4 (a) $f(b) = \sqrt{b}$, (b) $f(b) = \dfrac{1}{\pi} \arccos(1 - 2b^{3/2})$.

7.5 $g[n] = \frac{1}{2} f[n] + \frac{1}{4}(f[n-2] + f[n+2])$.

7.6 $\omega_c = \pi / N$.

7.7

0	1	0
−2	0	2
0	−1	0

7.8 $H(u, v) = -(u^2 + v^2)$.

7.9 The Wiener filter is proportional to $e^{-|t|}$ (in the time domain).

7.12

Moment no.	00	10	01	20	11	02
b	4	−1	0	13	9	15

7.13 Avocado.

7.14 $b_{kl} = c_{kl} - h_{kl}$ (for $k + l = 2$ or 3).

7.15 1.4.

Appendix A

A.3 (a) $\frac{4}{13}$, (b) $\frac{15}{52}$.

A.5

Event H	$\{1,2,3,7\}$	$\{4,8\}$	$\{5,9\}$	$\{6,10\}$
$p(H)$	0.4	0.2	0.2	0.2

A.6

(a)	0.1	0.2	0.3	0.4
(b)	0.1	0.3	0.6	
(c)	0.1	0.4	0.5	

A.7 $p(K_1 \mid R) = \frac{8}{23}$.
A.10 $E(\mathsf{X} + \mathsf{Y}) = 13.4$ ((a) and (b)),
$V(\mathsf{X} + \mathsf{Y}) = 127.8$ (a) and 164.04 (b).
A.13 $E(p) = \frac{1}{2}$, $V(p) = \frac{3}{4}$.
A.15 $P_3^c = P_3 - 3P_1P_2 + 2P_1^3$.
A.17 $C[\mathsf{m}, \mathsf{n}] = 3.75$.
A.21 $\hat{a} = \bar{n}/(\bar{n} + 1) = 0.8$.
A.23 (a) $\frac{1}{2}$ and $\frac{1}{12}$, (b) $1/a$ and $1/a^2$.
A.28 $(-\infty, -2.935)$.

APPENDIX D

Selected Further Reading

Image processing

K. R. Castleman, *Digital Image Processing* (Prentice Hall, 1979).
R. C. Gonzalez and P. Wintz, *Digital Image Processing* (Addison-Wesley, 1977).
P. Haberäcker, *Digitale Bildverarbeitung* (Hanser, 1989).
T. S. Huang (ed.), *Picture Processing and Digital Filtering* (Springer-Verlag, 1979).
A. K. Jain, *Fundamentals of Digital Image Processing* (Prentice Hall, 1989).
M. Kunt, *Traitement numérique des signaux* (Dunod, 1981).
R. Lewis, *Practical Digital Image Processing* (Ellis Horwood, 1990).
W. Niblack, *An Introduction to Digital Image Processing* (Prentice Hall, 1986).
W. K. Pratt, *Digital Image Processing* (Wiley, 1978).
A. Rosenfeld and A. C. Kak, *Digital Picture Processing* (Academic Press, 1982).

Signal processing

F. de Coulon, *Théorie et traitement des signaux* (Dunod, 1984).
A. V. Oppenheim and A. S. Willsky, *Signals and Systems* (Prentice Hall, 1983).
A. Papoulis, *Signal Analysis* (McGraw-Hill, 1977).
S. D. Stearns, *Digital Signal Analysis* (Hayden, 1975).

Information theory and coding

F. M. Reza, *An Introduction to Information Theory* (McGraw-Hill, 1961).
J. F. Young, *Information Theory* (Butterworth, 1971).

Recognition

J. T. Tou and R. C. Gonzalez, *Pattern Recognition Principles* (Addison-Wesley, 1974).

Geometry, algebra and functional analysis

J. Dieudonné, *Linear Algebra and Geometry* (Hermann, Paris, 1969).
J. Dieudonné, *Fondements de l'Analyse Moderne, Vol. I* (Gauthier-Villars, 1969).
T. Ewan Faulkner, *Projective Geometry* (Oliver and Boyd, 1960).

Astronomy and light detection

J. D. Kraus, *Radio Astronomy* (Cygnus-Quasar, 1986).
C. D. Mackay, Charge-coupled devices in astronomy, *Annual Review of Astronomy and Astrophysics*, **13**, 1986, p. 255).

Statistics

A. Hald, *Statistical Theory with Engineering Applications* (Wiley, 1962).
E. Kreyszig, *Introductory Mathematical Statistics* (Wiley, 1970).

Special algorithms

J. Högbom, Aperture synthesis with a non-regular distribution of interferometer baselines, *Astronomy and Astrophysics* (supplement), **15**, 1972, p. 417.
L. B. Lucy, An iterative technique for the rectification of observed distributions, *The Astronomical Journal*, **79**, 1974, p. 745.
W. H. Richardson, A Bayesian-based iterative method of image restorations, *Journal of the Optical Society of America*, **62**, 1972, p. 55.

APPENDIX E

Glossary

Aliasing: Erroneous identification of high-frequency components with low-frequency ones.
Alternative: See *hypothesis*.
Amplitude: The maximum value of a harmonic signal.
Amplitude spectrum: The moduli of a (complex) Fourier transform.
Analog signal: Continuous signal; the limiting case of a digital signal, for which the digital time intervals are small and for which the signal values do not change appreciably.
Autocorrelation: Process intended to detect periodicities in a signal.
Bandwidth: Quantity measuring the frequency domain extent of a system.
Bit: Abbreviation of *binary digit*: 0 or 1.
CCD chip: Two-dimensional integrated (semiconductor) light detector. Characterized by high quantum efficiency and proportionality between incoming and detected amount of light.
Centre of gravity: Point defined by having coordinates equal to the two first-order distribution moments.
Charge coupled device: see *CCD chip*.
Cluster: Group of objects with the same identification; see *recognition*.
Coding: Process in which a set of data is converted into new (often binary) data, notably for purposes of reducing storage or transmission demands.
Contrast: Measure of the distribution of grey levels in an image.
Convolution: Linear mathematical process (see p. 91). Replaced by multiplication in the frequency domain (after z- or Fourier transformation).
Correlation, cross-correlation: Process quantifying the similarities between two signals; see also *autocorrelation*.
Cross ratio: Quantity conserved under central projection (Section 6.4).
Deciphering: See *decoding*.
Decoding: The conversion of a code into the information which generated the code.
Deconvolution: Determination of a signal the convolution of which with another, known, signal results in a third signal.
Delta signal: Special signal, equal to 0 for all values of time except one, where it equals 1.
Detection: Process intended to establish the presence of certain objects or signal components in an image.

Digital: Numeric; specified by means of digits (numbers). Also refers to *discontinuous* signals.

Digital Fourier transformation: The expression of a signal as a linear combination of signals from the harmonic orthogonal set.

Digital image: Image characterized by numbers. Specified by means of a rectangular array of numbers; the numbers indicate grey level or light intensity, while the positions correspond to those of the physical image.

Digital time and digital position: Variables for enumerating the values of a digital signal or a digital image.

Digitization: Process intended to represent an analog signal as a digital signal.

Distribution: Statistical characterization of a stochastic variable.

Energy: The sum of squares of the values of a signal.

Energy spectrum: The squared moduli of a (complex) Fourier transform.

Entropy: Measure of the degree of randomness or smoothness in a statistical distribution.

Estimation: Used here in the context of statistical distribution parameters.

Estimator: Stochastic variable used in estimation.

Expectation: See *mean*.

Exponential signal: Signal of form z^n, where z is a complex number and n the digital time.

Fast Fourier transform: Algorithm minimizing the computing time for a DFT.

Feature: Image parameter; quantity characterizing an image or its objects.

Feature space: The range of variation for one or more image parameters.

Filtering: System effect specified in the frequency domain.

Flat-fielding: Process correcting differences in sensitivity between the individual pixels of a CCD chip.

Fourier series: Representation of a signal (notably a periodic one) as an infinite linear combination of harmonic signals.

Fourier transformation: Representation of a transient analog signal as an integral over harmonic signals. The coefficients of the individual components are called the *Fourier transform* of the signal.

Frequency: The number of oscillations per second, in particular for a periodic signal. Also used to denote statistical occurrences.

Gaussian distribution: See definition of *normal distribution* in Section A.9.

Geometric decalibration: Correction for geometric distortion.

Grey level: Measure of intensity within a black/white image.

Harmonic: Sinusoidal.

Harmonic orthogonal set: Set of exponential basis signals, expressed as powers of the complex unit roots.

Histogram: Graphical means for specification of (statistical) frequencies. Approximates an exact statistical distribution.

Histogram equalization: Process used for contrast enhancement.

Hypothesis: Tentative specification of a statistical distribution, the validity of which is tested by experiment. Should normally be accompanied by an *alternative* hypothesis.

Image enhancement: Image processing method altering the histogram of an image towards a broader level distribution.

Image restoration: Correction for image degradation typified by blurring or smearing.

Image quality: Subjective measure for the information content in an image.

Impulse response: Output corresponding to a delta signal input. Called *point spread function* in the image case.

Independence: For two or more stochastic variables, the property that the outcome of one does not affect the outcome of the others.

Information content: May, for example, be quantified by a lower limit to the transmission resources needed.

Input: Signal (image) to be processed by a transformation.

Intensity: Measure of physical light energy per unit area and unit time.

Interpolation: Method for reconstruction of an analog signal from its samples.

Invariance: Transformation property expressing independence of time or position.

Invariant moments: Distribution moments, calculated in a certain coordinate system determined by the distribution itself.

Inverse filtering: The elementary process for image restoration.

Least squares approximation: Optimal approximation according to signal distance.

Level: See *grey level.*

Light quantum: See *photon.*

Line spread function: The output of an image processing system for an input line (and nothing else); cf. *point spread function.*

Maximum likelihood estimation: Estimation method in which the parameter to be estimated is required to maximize the probability (likelihood) of the actual observations.

Mean: The constant which, in the least squares sense, is the best approximation to a stochastic variable.

Moment: Distribution parameter; the mean of a power function.

Noise: Irrelevant signal superimposed upon a given (information-carrying) signal.

Norm: For a signal, the signal distance from the zero signal.

Normal distribution: See definition in Section A.9.

Normalization: Division of a signal by its norm, so that the result is of norm 1. *For probability distribution*: Division of a set of positive numbers by their sum, so that a probability vector results.

Nyquist frequency: Half the sampling frequency; the aliasing limit.

Orthogonality: Two signals are said to be orthogonal if their scalar product is 0. If, moreover, they are *normalized*, they are known as *orthonormal.*

Output: Exit signal; the result of a signal transformation.

Oversampling: See *sampling.*

Parameter: Fixed value of digital time. Also: quantity characterizing a distribution or an image. An *image parameter* is, in particular in the context of recognition, also known as a *feature.*

Parameter space: See *feature space.*

Photon: The smallest unit of light.

Pixel: See p. 10.

Planck's constant: See p. 11.

Point spread function: The two-dimensional impulse response, that is, the system image of a point source.

Poisson distribution: Statistical distribution describing, for example, the arrival of uncorrelated photons.

Power: Energy per unit time.

Probability: The relative frequency with which an event occurs, in a large number of repeated, identical experiments.

Probability density: Continuous probability distribution.

Probability distribution, probability vector: Specification of the probabilities of all possible outcomes or events associated with an experiment.

Projection: Process ensuring a least squares approximation.

PSF: see *point spread function*.

Quantization: Process in which groups of similar values are replaced by one representative value for each group.

Quantum efficiency: For a detector, the ratio between the number of detected photons and the number physically incident upon the detector.

Recognition: Process intended to identify a given image with a reference image.

Rectangular distribution: See *uniform distribution*.

Rectangular signal: Signal of constant value in a finite interval, being zero outside this interval.

Redundancy: Excess information, the purpose of which could be elimination of noise.

Resampling: Sampling in a new set of points in a formerly interpolated signal.

Resolution: The lower limit for a physical system as regards separation of nearby time values or signal values.

Ringing: Periodic image pattern produced by filters with sharp edges.

Sampling: Conversion of an analog signal into a digital one consisting of representative values. *Undersampling (oversampling)*: sampling with too few (too many) representative values as compared to a satisfactory sampling.

Sampling theorem: Theorem according to which a sum of harmonic signals can be reconstructed uniquely from its samples, provided all frequencies fall below the *Nyquist frequency*.

Scalar product: The product sum of the components of two signals. (The second signal should, if complex, be conjugated.)

Segmentation: Division of an image into smaller, connected areas, often corresponding to physical objects.

Sensitivity: See *quantum efficiency*.

Separability: A digital image is separable if it is a product of two one-dimensional signals.

Signal: The physical mediator of information; function (in the mathematical sense).

Signal distance: The square root of the sum of squares of differences between the components of two signals.

Spectrum: Distribution over frequencies or wavelengths, for example, for light or other types of electromagnetic radiation. Is evaluated by Fourier transformation or spectral estimation. See also *amplitude spectrum* and *energy spectrum*.

Statistic: Stochastic variable determining the result of a test.

Step signal: Signal of constant value in an infinite interval.

Stochastic process: Signal consisting of a stochastic variable for each digital time value.

Stochastic variable: Function which transforms or generates probability distributions.

Surface brightness: The fundamental physical quantity, the image counterpart of which is image *intensity*, expressed as a *grey level* or a photon number.

System: Temporally (or spatially) invariant transformation.

System function: *z*- or *Fourier* transform of the impulse response of a system.

Template: Simplified reference object used for recognition.

Test: Statistical method for the accepting or rejecting of a *hypothesis*.

Time invariance: See *invariance*.

Transformation: Signal change preserving linear combinations.

Transient: Signal which differs from 0 only on a bounded time interval.

Translation: Transformation equivalent to a shift of time (or position).

Undersampling: See *sampling*.

Uniform distribution: Distribution consisting of equal probabilities.

Unit delay (element): Translation by one digital time unit.

Unit impulse: See *delta signal*.

Variance: For a stochastic variable, its mean signal distance squared from its expectation value.

Variance equalization: Method for image enhancement based upon the level distribution variance as a measure of contrast.

White noise: Stochastic process of constant power spectrum.

Wiener deconvolution: *Inverse filtering* followed by *Wiener filtering*.

Wiener filter: Optimal filter which, for given spectral properties of signal and noise, minimizes the signal distance between input and output.

***z*-transformation:** The summation of a power series, the coefficients of which equal the components of a given digital signal.

Index